Routledge Revivals

Beyond Physics

Originally published in 1930, Sir Oliver Lodge proposes a connection between physics and philosophy, or as he describes it, a key to unlock the intricate connection between mind and matter. A response to early twentieth century mathematically-led philosophy, Lodge looks at physics from a physical direction rather than from a theoretical model. This title will be of interest to students of philosophy as well physics.

I08941:34

Beyond Physics

Or the Idealisation of Mechanism

Sir Oliver Lodge

Routledge
Taylor & Francis Group

First published in 1930
by George Allen & Unwin, Ltd.

This edition first published in 2016 by Routledge
2 Park Square, Milton Park, Abingdon, Oxon, OX14 4RN
and by Routledge
711 Third Avenue, New York, NY 10017

Routledge is an imprint of the Taylor & Francis Group, an informa business

Publisher's Note

The publisher has gone to great lengths to ensure the quality of this reprint but points out that some imperfections in the original copies may be apparent.

Disclaimer

The publisher has made every effort to trace copyright holders and welcomes correspondence from those they have been unable to contact.

A Library of Congress record exists under LC control number: 30031683

ISBN 13: 978-1-138-19164-8 (hbk)
ISBN 13: 978-1-315-64036-5 (ebk)
ISBN 13: 978-1-138-19165-5 (pbk)

BEYOND PHYSICS

OR THE IDEALISATION OF MECHANISM

BY

SIR OLIVER LODGE

D.Sc., LL.D., F.R.S.

Being a survey and attempted extension of Modern Physics in a philosophical and psychical direction

LONDON

GEORGE ALLEN & UNWIN LTD

MUSEUM STREET

FIRST PUBLISHED IN MAY 1930
SECOND EDITION AUGUST 1930
THIRD IMPRESSION (SECOND EDITION) JANUARY 1931

PRINTED IN GREAT BRITAIN BY
UNWIN BROTHERS LTD., WOKING

PREFACE

MANY times it must have been urged that physics and psychics were interlocked, at least to the extent that probably every psychic event had a physical concomitant, but hitherto no attempt has been made to imagine a mechanism or physical process for this interaction. The ether of space has not been a medium much attended to in philosophy; and if the ether contains the key to the connexion between mind and matter, that key has been effectively hidden. Groping among the intricacies of modern physics I have found something that feels like a key; it has a handle at one end; what it has at the other end is still rather speculative. A good deal of cleaning up and brightening is needed before we can seriously try if it will fit the lock; and there is always the possibility that the handle may give way. But if we were too elaborately cautious about trying things, we should not make progress except in easily verifiable directions. The enterprise is worth some risk.

Such verification as has so far been forthcoming takes the form of a general approval of the broad outline, from those with whom I am in touch under conditions different from those of every day, and whose range of experience is rather more enlarged than ours. I do not emphasise this general approval, nor in this book do I say anything about that side of experience. Whether my main contention has any survival value for philosophy must depend on the direction in which truth lies.

Even modern physics is not so unalterably certain that it can be incorporated into philosophy without a qualm; yet many philosophers are trying their hand at some kind of incorporation. The whole subject is full of living interest; and a guiding even if also a groping hand may not be unwelcome in a tangled region in which beaten tracks are scarce and where the light of day can hardly penetrate the forest of abstract and uninterpreted symbols. I have long written round about this subject of mind and matter in what is called a popular manner: I now attempt a slightly more ambitious treatment, so as to appeal more directly to experts. Whether I succeed in the task of pacifying physicists and interesting philosophers, I at least contribute an intelligible survey and criticism of certain features in modern physics, which I hope will prove useful.

As for the main contention of the book, I have repeated it in various forms, for I believe that in the main it is right. Details will certainly have to be amended as knowledge grows, but I would request critics to take the whole argument as seriously as they can, and to give it the benefit of the doubt.

The sub-title of this book, "The Idealisation of

Mechanism," is liable to be misunderstood; as if I had written, "The Apotheosis of Mechanism." Nothing of the kind! My aim is not the glorification of mechanism, but its rational interpretation and understanding. Every piece of man-made automatic machinery is the outcome of mental ingenuity, it is saturated with design and purpose. The recognition of that fact constitutes its idealistic interpretation.

Machinery can be regarded solely from the mechanical point of view—its *modus operandi* can be studied, its intricate behaviour enjoyed. An intelligent child can be interested in, for instance, a gramophone which needs no attention and changes its own records, or in any other device not too complex and dangerous for a child to watch in operation. The child will delight in movements which simulate intelligence—they seem to awaken a fellow-feeling—but he is unlikely to give thought to the mental processes involved in its construction; he will not think of the designer or of the successive improvements through which such mechanical perfection was gradually evolved from lowly beginnings. In other words, a child will not regard mechanism from the idealistic point of view: he will be satisfied with its "behaviour". A professed engineer or artificer, on the other hand, will be impressed with the ingenuity of the inventor and designer; the mental aspect will be dominant from his point of view. Yet he need not pour scorn upon the simpler kind of appreciation; he can share the child's delight in the detailed working of a perfect machine, while still fully aware that there is far more to be said, much elaborate truth which the child could hardly understand. Something of value reveals itself to either kind of contemplation, neither by itself is complete.

Aye, and there is yet a third point of view, that of the Poet and the Painter—one from which neither the physical working nor the psychological implication is specially attended to, but one from which, as the Poet Laureate has recently emphasised, we can revel in the super-added beneficent gift of beauty. For, as Dr. Mozley said long ago, "Nature in the very act of labouring as a machine also sleeps as a picture".

This goes beyond my parable of the child and the machine; yet if the bearing even of this parable were fully grasped, if it could be recognised that no one statement can be comprehensive enough to exhaust the whole truth about anything, many of our semi-scientific semi-theological controversies would evaporate into thin air. Very well, this book seeks to sustain and even elaborate the physical side of things, and yet to insist on the domination of the spiritual. *Mens agitat molem, et magno se corpore miscet.*

OLIVER LODGE

Christmas, 1929

CONTENTS

BEYOND PHYSICS

CHAPTER I

THE PHILOSOPHY OF SCIENCE

Many attempts have been made to reconcile the advances of recent years in scientific knowledge with the outlook of philosophy. Scientific men who have explored the mechanism of the universe have mostly adopted a mechanistic or materialistic view, in the hope that a mechanical theory can be complete in itself and is already approaching completion. But although they have made progress in this direction and taught us a good deal about the machinery, they have been pulled up sooner or later by the troublesome phenomenon of consciousness, and by the interpretation of mind generally. So some of them have been led into a deep-seated scepticism as to whether experience gives us a true interpretation of the universe, and whether our scientific theories are more than a convenient method of formulating and arranging the facts. They have not found the idealistic point of view, taking mental experience as a basis, easy to reconcile with the accurate metrical determinations of science; still less with the conception, prevalent a few years ago, that the universe is a coherent self-contained well-organised automatic machine, upon which mental phenomena are grafted as an illusory epiphenomenon—a fruit of the behaviour of complex molecules. That it is a machine, up to a point, is true enough, but it is a thoughtful machine, and therein lies the puzzle; for in that respect it is quite unlike any other machine. We have gradually found that we cannot make a working model even of the ether: we certainly cannot make a model of a thinking and planning live thing. Those who attempt it probably hold that cause and effect are out of date, and ought to be replaced by a mere perception of sequence; that the adaptation of means to ends, or of an organism to its environment, is a consequence of the conditions of survival and heredity, and furnishes no trustworthy indication of design and purpose. For although the fact of evolution is admitted by all, it can be argued that all that we really observe is change, automatic and inevitable adjustment to environment, without any necessary increase in value. What we study in science, it is said, are things that can be weighed and measured, about which we can lay down quantitative laws; and it is hoped that as science progresses we shall more and more be able to reduce everything to a mechanical and inevitable sequence, representing the action of one material body on another, after the fashion of the Newtonian system of astronomy.

On the other hand, those who are impressed with the mental or idealistic foundation of the universe find it hard to realise how life and mind have entered into relation with matter, or what the materialistic and essentially mechanical system really means, from the mental point of view. There is, as it were, a conflict of opposites, which has not been resolved. The notion that evolution is a self-sufficient automatic process, without guidance or meaning, has proved repugnant to a great number of simple people, whose intuitions are against it; and in default of a reconciliation they are unwisely prepared to disbelieve and discard the fact of evolution itself. This last, from any point of view, is a hopeless attitude. But the admitted difficulty has led some to doubt the testimony of experience, and to take refuge in a mystical or theological view, which leaves everything unexplained. For that would be as if we said that things are as they are because of a creative Fiat, and that exploration of any gradual process by which things acquired their properties verges on impiety. The apprehended facts of nature and the intuitive spirit of man are thus liable to seem inharmonious, so that the upholders of either side tend to ignore the other.

Those who would regard causation as sequence, and evolution as mere change, are on one side: they have made great progress in their department, and are apt to consider a more idealistic view as illegitimate and unscientific. Science is measurement, they say, and anything that cannot be reduced, or shows no sign of being reducible, to quantitative and mathematical formulæ, is unworthy of attention. This doctrine, occasionally held by physicists and mathematicians, is a natural outcome of the brilliant successes in that department of knowledge, and is beginning to penetrate through chemists to biologists, not so much in connexion with their present achievements, but in respect to their future ambitions. The famous *Utinam* of Sir Isaac Newton, in which he set forth as the ideal aim of science the reduction of all phenomena to the interaction of the forces between particles, may be pleaded in their favour.[1] At the same time it must be remembered that whatever view is held by the mathematicians of to-day, the great originators of the scheme, from Newton downwards, were not satisfied by it. They felt instinctively that it was not the whole truth, and that there were deeper and higher modes of regarding existence. The

[1] "Would that the rest of the phenomena of nature could be deduced by a like kind of reasoning from mechanical principles. For many circumstances lead me to suspect that all these phenomena may depend upon certain forces, in virtue of which the particles of bodies, by causes not yet known, are either mutually impelled against one another, and cohere into regular figures, or repel and recede from one another; which forces being unknown, philosophers have as yet explored nature in vain."—(From Newton's *Principia*.)

greatest Natural Philosophers have always tried to look at the world, not only in minute and separate detail, but as a whole; they have trusted experience to lead them ultimately in the direction, not merely of convenience of statement, but actually of genuine truth.

There must surely be (we would all say) some connexion or rapport between the mind of man, on the one hand, and the facts of nature, on the other. It has even been held that our aspect and interpretation of the universe is a construction of the mind of man, and that what the real nature of existence is, apart from its apprehension by human mind—what things are in themselves—must be for ever unknowable. Thus scepticism obtrudes itself, sooner or later, in every direction. Some are sceptical about the mechanism, while others are sceptical about any meaning or purpose underlying it. The reconciliation and unification of these opposing tendencies, and the inclusion of the whole in a wider generalisation, must surely be the end and aim of modern philosophy as well as of modern science. Evolutionists, however much they may discard the idea of growing value and teleology and entelechy, can hardly fail to recognise that there is not only change but improvement. The horse, for instance, is a more splendid animal than *eohippus* and its other ancestors. Species do progress from lower to higher forms. And that admission, if made, does suggest that the whole is advancing towards some ultimate perfection, and that the stages in the process cannot be devoid of meaning and purpose.

Meanwhile, the origin of species remains essentially an unsolved problem. The old idea of special creation, each creature brought into being complete as it is, is no solution at all: evolutional creation is quite unlike Milton's or Haydn's representation; it is gradual, and the process by which creatures attained their present forms can be scrutinised by palæontological science. But whether the evolution of organisms is a continuum, out of which the intermediate forms have dropped by extinction (so that the chain of evolution is full of missing links, and existing species are the fragments that remain), or whether mutations took place by sudden jumps, which were subsequently inherited and, if favourable, survived, so that there are no missing links, are problems still open to debate. Experience shows that variation, heredity, adaptation to environment, struggle for existence, and survival of the fittest, are *veræ causæ*. But how variations arose, what makes them heritable, whether adaptation can be secured by efforts of the individual, and whether there is any guiding principle underlying the whole process, are questions still unanswered. To advance at all, to emerge from blank and fruitless scepticism, to proceed in the absence of certain knowledge, requires some act of faith—faith, let us say, in the growing value of existence as such, in the power of the human mind to appreciate real truth, and in the value of experience as a guide to reality. Moreover, by "experience" must be meant, not only the direct evidence of the senses, but the inferences and intuitions, the theories and working hypotheses, that are based upon and enlarge and expand direct sensory apprehensions. We have experience, not only of material things, but of mental states also; and no part of experience should be neglected if we are to make an attempt to comprehend the whole.

As for myself, I am supposed to be tarred with the brush of unorthodox researches into psychic and psycho-physical phenomena. The only brush, however, with which I admit being tarred is a brush which has been dipped into facts of experience. I admit that they are facts at present unadmitted by the majority of my colleagues, but a minority have assured themselves that they are facts; and if so they cannot fail to be of the utmost importance in any complete or philosophic scheme. On that basis I make an incipient attempt to unify physics and psychics; and towards this unification I wish to emphasise the importance of the universal connecting medium, the ether of space. I believe that this substance or sub-stantial entity will ultimately be found to be of the first importance both in science and in philosophy; I believe that it will act as an instrument of unification between mechanism, on the one hand, and spiritual guidance, on the other. If the ether is a substance of universal prevalence—as in physics it appears to be—then it may be the real vehicle of mind and spirit. If so, then it must be by or through the process of what we call incarnation—a connexion between ether and matter which has still to be understood—that the undifferentiated mind develops into separate personality. It proceeds by utilising the essential discontinuity of matter to partition itself off into free and independent units, so that in association with matter they may acquire an individuality of their own, and thus by free and personal development enhance the value and complexity of the whole.

If mind always requires a physical vehicle (as our experience seems to show that it does), then, since a material form is inadequate, the Absolute Mind, whether differentiated or not, must exist in the continuous ether; for that is the only physical entity with perfect properties known to us. Mind transcends the ether, yes, but utilises it as its instrument and vehicle, as will be argued later. In itself, and apart from the evolutionary processes conducted with the aid of matter, it would be self-sufficient all-inclusive solipsistic Mind; but it can blossom or evolve into separate personalities by the device of incarnation; employing for that screening and individualising process the special modifications of ether that make direct appeal to the bodily senses of men and animals, and that we apprehend as the universe of matter. This it is which has formed the subject-matter of science as at present understood; and we have recently learned to emphasise afresh every essential discontinuity. Matter is known to consist of detached particles linked together into a cosmos by unexplained and non-sensuous forces, either in what may be called space, or in a physical but non-material continuous perfect substance—that is, one without any of the imperfections which dissipate energy and are inevitable in any molecular aggregate—which at the same time is the seat of all potential energy. And now matter itself is turning out to be a form of energy likewise, so that the whole physical universe is being resolved into ether and energy, which last may be a special variety of motion. In what way this omnipresent entity is able to fulfil the behests of life and mind has not yet even been imagined; but if the ether exists, it is unlikely that life and mind have not made use of it; and it is in some such way that I would seek to fathom the meaning and mystery of existence as we know it.

I am sure that the explanation is not to be found in a study of matter alone, nor of mind alone, nor of both together, without an intermediary third.

Meanwhile, our dual nature is fairly obvious, our mind is certainly connected with our material organism. Matter, as Bergson has hinted, is a screen and an obstruction, a useful because an individualising screen isolating us from the rest of the psychical cosmos, a limitation of that which in itself has a far wider and more permanent scope. My attention has been called by Mr. F. C. Constable to a sentence of Kant which corresponds closely with my own view, and with it I will open the next chapter.

CHAPTER II

THE RELATION OF MATTER TO LIFE AND MIND

"The body would, in this view of the question, be regarded, not as the cause of thought, but merely as its restrictive condition, as promotive of the sensuous and animal, as but a hindrance to the pure and spiritual life; and the dependence of the animal life on the constitution of the body would not prove that the *whole* life of man was also dependent on the state of the organism. . . . If we could intuit ourselves and other things as they really are, we should see ourselves in a world of spiritual natures, our connection with which did not begin at our birth, and will not cease with the destruction of the body."—KANT's *Dialectic*, Meiklejohn's translation, pp. 473–4.

To treat of the relation between life and matter, after the problem has been discussed from so many points of view and in the light of the teachings of philosophers of all ages, must seem perhaps ambitious and certainly presumptuous; but I presume that philosophers do not consider that they have completely solved the agelong problem, and some may be willing to hear what a physicist has to suggest in the light of our present knowledge of nature. I have no wish to challenge controversy, but controversial topics cannot be wholly avoided. For instance, throughout I shall speak of the interaction of life and matter, as if they were two distinct though inter-connected things; and yet I am well aware that this word "interaction" itself strikes a controversial note. The theories of Epiphenomenalism and Parallelism are in the field against it; though as to the question how far they are still supposed to hold that field there may be differences of opinion. Neither of those views, however, in the least commends itself to me, while the term "interaction" does; though I may admit from the outset that interaction of two distinct things cannot be the last word, and that an ultimate unification is surely the end and aim of Philosophy.

Meanwhile, whatever may be ultimately accomplished, we have not enough information at present to treat of mind and matter as *one*, in any satisfactory and scientific sense, and so for the present we must deal with them as at least different aspects of something unknown, and admit the obvious fact that they interact.

Between life and mind, however, I do not propose to draw any radical distinction; though mind is conscious of itself, and life presumably for the most part is not. I regard life as the rudiment of mind, and mind as the conscious part of life: they are, so to speak, the same thing in different stages of development. I do not regard either as a mere consequence of the complexity of material molecular organisation; though it is true that in our present terrestrial condition, with direct outlook limited by animal bodies and animal senses, a material organisation is necessary for our apprehension of life and mind; they must be incorporated in or associated with a material form if we are to perceive them. We are acquainted with life by our ability to see and touch living organisms; we are acquainted with mind and consciousness in a primary and direct manner by our own experience; and we attribute the same sort of experience to the bodies of similar type which our senses tell us exist around us.

It is only through primitive experience that we have become consciously acquainted with matter. Matter itself is an inference. The whole external world is an inference, based on our sensations. And life and mind, regarded intellectually as objective, are further inferences, not based directly on our sensations, but developed from them with a large mental contribution. It is not by direct apprehension, but by mental contribution, that we become acquainted with works of Art, whether in the form of literature, or poetry, or music, or painting, or architecture and sculpture. The physical basis of what we call a work of art is simple enough, and can be apprehended by the animals. The artistic aspect, the whole meaning and value of such things, depends on our human interpretation; that is to say on mental faculties, highly developed in some individuals, and only dimly exercised by the majority. The physical basis, alone, is given by our senses: all the rest belongs to the ideal world of thought and imagination. A work of art must be constructed by the muscles, but it cannot be explained in terms of muscular action. A cathedral must be built by quarrymen and stonemasons, but the design is not theirs: it was conceived in a mind which perhaps never touched a stone. Yet the design itself had to be displayed by muscular action, even if only by the movements of a pencil. A poet or a musician has to make black marks on paper, or use an instrument, in order to display or incarnate his ideas in matter, but no one supposes that his muscular activity explains the poem or the symphony. That, in its essence, does not belong to the material world. Nevertheless it has to interact with the material world before others can perceive or realise it.

Hence the word "interaction" seems necessary and satisfactory as far as it goes. What is incorporated in matter is not the idea itself, but an incarnation of the idea, a material representation which is sufficient to call out or evoke similar ideas in other minds capable of appreciating it; provided they are possessed of the necessary sense organs for the apprehension of the material instrument by which it is displayed. In other words, matter is an instrument in which thought can be incarnated and manifested. All our conscious actions are thus representations or manifestations of our thoughts: and without a bodily organism we could neither manifest our own thoughts nor normally apprehend those of others.

Our bodily mechanism consists not only of muscles, through which alone we act on the external world, it also contains a brain and nervous system which controls and works those muscles and receives impressions from our sense organs. The brain is, as it were, the primary instrument which mind utilises, and through which all the rest is accomplished. If the brain is damaged, or out of order, the manifestation is imperfect, or may cease altogether; and this familiar fact has led some people to postulate that mind has no existence apart from brain, that brain is not so much the instrument of mind as mind itself, and that when the brain is destroyed the mind is destroyed too.

I urge that this does not follow at all, and that it is contrary to all analogy. A close examination of the brain will not explain thought, though it will show us the mechanism by which thought is reproduced in perceptible material form. Examination of the instruments of an orchestra, or the strings of a piano, would never yield a symphony or a sonata; and yet those instruments are necessary for its reproduction or manifestation. A savage wandering in the interior of an organ would be no nearer the understanding of music; nor would he be destroying music if he wielded a hatchet in his journey, though he would be injuring its presentation. Similarly, even if we could see the molecular processes going on in a brain, the rhythm would be interesting, but we should not be any more enlightened than if we merely witnessed the movements of conductor and violinists in an orchestra. The printed words in a volume of Shakespeare would mean nothing to a savage, however full of human nature they are perceived to be by a competent and understanding mind. A printer's fount is a receptacle of potential sense, but a mental operation must accompany the sorting out.

If all this, or something like it when properly expressed, is generally admitted as representing something of the truth at the level of mind, let us apply similar considerations to the lower forms of life. A seed or a germinal vesicle is in itself a molecular aggregate consisting of an enormous number of atoms, which themselves are now known to be each a set of minute electrical particles revolving round an electrified nucleus. These atoms have grouped themselves into molecules of considerable chemical complexity; and some of these molecules, under a strange impulse, have constituted a substance known as protoplasm, which somehow becomes the seat of vital processes, and demonstrates vitality. If this protoplasm is interfered with in a drastic manner, it may show no sign of life; whereas if it is preserved intact, the seed is able to germinate and bud, assimilating molecules and energy from the rest of the material world, so as to build up the complex, elaborate, and perhaps beautiful structure of a plant or an animal; and it can continue the process without limit for innumerable generations. But no examination of the seed or germ will explain its vitality. Associated with or incorporated in it is something which not only enables it to accrete into its own structure otherwise alien material, but also exercises specific control over that material, building it up into definite localised forms of specific type—a type which does not depend on the material, but is entirely dependent on the indwelling specific essence, of which the material is only the vehicle and demonstration.

It is easy enough to interfere with and destroy vitality, that is to say the manifestation of life: we can assist it to flourish, or we can retard it; but we have no other control over it, and no real understanding. The essence of life is beyond us: we know not whence it comes, nor whither it goes. So far as we know at present, there is no life without antecedent life. Life is passed on from one organism to another, and can increase illimitably. It is dependent on matter and physical and chemical energy for its manifestation and development; but neither matter nor energy can explain the thing itself, or give any idea of its marvellous properties.

I call the properties of life "marvellous" with due appreciation of the meaning of the term. For surely the familiar fact that an acorn contains within itself the power to produce a whole forest of oaks; and still more, perhaps, the equally familiar fact that a hen's egg kept warm for a few weeks, though at first apparently a mere mass of unformed protoplasmic material, can result in a fledged creature, with bones and muscles and nervous system and eyes—which can emerge and fend for itself, stand and peck with discrimination, though perhaps hatched out in a mere incubator—is as marvellous as anything we can think of, if we were not so accustomed to it as to lose the sense of wonder. And even the movements of a lowly protozoon or an amœba, as it crawls and absorbs nutriment and excretes and grows and subdivides and multiplies and continues, is more than anything we are able to account for in terms of the properties of matter.

Yet matter displays these properties: animated matter is vitalised. Vital functions are not natural to inert matter, merely subject to the ordinary laws of mechanics; so I argue that vitality must be due to the interaction of something which is not matter, but which utilises matter for its manifestation. What can be said on the other side? Many of the organic compounds found in living organisms, or secreted by them, have now been made in the laboratory; beginning with urea, and continuing up to sugar and starch and numerous other compounds. And it is sometimes said by students of organic chemistry, and by the bio-chemists who study protoplasm, that if we could contrive in the laboratory to extend the manufacture of these organic compounds until we had made a mass of protoplasm, and were able to subject it to suitable treatment, they would expect that artificial protoplasm to exhibit vitality and to manifest one or other of the signs of life.

Now I would not in the least seek to deny that proposition, I certainly would not contend against it; I would even regard it as reasonable. Indeed, from some points of view, the evolution of live creatures from inorganic material must be regarded as inevitable; because of the undoubted facts, (1) that living things have appeared on this planet, and (2) that this planet was at one time a mass of molten material, or even glowing gas, in which any display of life as we know it was impossible.

Hence it would appear that something of the kind must have occurred in the past; and what has happened in the past may be going on in the present, and may possibly be understood and controlled and humanly managed in the future.

But what then? Many would feel afraid of such a conclusion, and think that a self-acting mechanism of that kind would remove from the universe the need for a planning and creative Mind, so as to be out of harmony with certain deeply implanted instincts and religious ideas. Their fears seem to me groundless, and to be based on the mistaken term "self-acting". For the process we have assumed as possible (not yet, but some day possible), in a laboratory, is surely not a self-acting process at all. A chemist who in the future may discover how to construct protoplasm and to infuse it with vitality, is himself no self-acting machine. He surely is full of knowledge and contrivance and planning, and is conducting operations full of understanding and design. That life, therefore, when it appears, will not have come into being without antecedent life. The chemist or the physicist who has done it was alive, and has only succeeded in designing and accomplishing his task through the agency of a powerful mind. The phenomenon will not have occurred haphazard and without thought; there is nothing in the prospective achievement to which exception need be taken. Rather it might be welcomed, even by religious people, as showing what an amount of thought was necessary to produce any imitation of what actually now is. If we are wise we would never be afraid of any progress in knowledge, we would never oppose or obstruct the achievements of science, or, like Canute, try to set a limit to its advance with a "Thus far, and no farther!"

Analogies to Life

Meanwhile, returning to the point of view of Physics—and speaking as a physicist who hopes that one day life will be better understood, and some comprehension be attained of the way in which life and mind interact with matter—let us ask if there is any analogy, at present known to us, which is at all akin to the production, increase, and dissemination of life. I have elsewhere before now hinted that such an analogy exists in the phenomenon of magnetism; and, since the majority of people are necessarily not physicists, it is permissible and perhaps necessary to develop this idea and to elaborate the analogy, so far as it goes, by explaining what I mean; remembering always that an analogy is only an analogy, and that sooner or later it must break down, else it would be no analogy but the thing itself. Let me therefore for a few minutes talk in elementary fashion in terms of modern physics.

I will speak first on the relation between ELECTRICITY and MAGNETISM.

So long as an electric charge is at rest, or static, there is no connexion between Electricity and Magnetism: the connexion becomes apparent only when it moves. A charge in motion constitutes an electric current, and every electric current has magnetic properties. Its path is surrounded by circular lines of magnetic force as a curtain-pole is surrounded by rings: and if an electron is revolving in an orbit, the plane of that orbit constitutes a sort of disc magnet, like a plate of steel magnetised with north magnetism on one face and south on the other, sometimes called "a magnetic shell". A single electron revolving in a very small but specified orbit constitutes in fact the unit of magnetism, and is called a magneton. We find that every magnet is built up of magnetons, just as every electric charge is built up of electrons: and gradually it has been found that that is all that magnetism is. Every magnet, whether it be a permanent steel magnet or an electromagnet, is constituted of revolving electrons; it is really an assemblage of circular currents arranged with a certain amount of order, so as to face more or less in one direction.

These currents, when not artificially made, are excessively small in size, comparable in fact to atoms, and their number is enormous. When these molecular currents are arranged at random, mixed up in all directions, the substance is magnetisable, but not magnetised: in a magnet the currents are marshalled in ranks, so that a number of them face in one direction. So long as they are held in that direction, the substance continues to be a magnet; but when released they swing back again, and the magnetism apparently ceases. Some substances, like steel, are able to retain their molecules in the strained condition, facing more or less one way; and that is a permanent magnet. Hammering or heating enables them to settle back again, and so demagnetises the steel. Any substance in which the molecules were *all* facing in one direction would be saturated with magnetism; it could hold no more.

When a piece of steel or anything else is demagnetised, you are not to think of the revolving electrons as stopping, or of the magnetism as going out of existence; for that never happens, any more than an electron goes out of existence. There is a rearrangement of the particles, so that they produce no outside or perceptible effect. The magnetism can go out of our ken, but the electric charges are still there, and so are the molecular currents, only they are no longer arranged so that their lines of force co-operate—spreading out into space so that they can be displayed and experimented upon; they are, as it were, shut up among themselves, and are as unreadable as a shut book.

The act of electrification is a sorting operation: a process of separating the particles, accumulating the positive in one place, the negative in another; the region between being filled with lines of force, called an electric field. Similarly, the act of magnetisation is a drilling operation: a process of organising and facing the molecular currents round, so as to have an orderly arrangement. The whole neighbourhood then becomes a magnetic field, filled with magnetic lines of force. But the magnetism has not been generated, it has only been displayed. In the same sense an electric charge has never been generated by any of our operations, it has only been made manifest.

Generation and Manifestation

It is important even at this stage therefore to discriminate between things that can be generated and things that were pre-existent and can only be exhibited. Sound, for instance, can be generated. But then what is sound? It is merely a vibration; so that the generation of sound is essentially the causing of pre-existing things to vibrate, as when you bow a string or strike a bell. Heat, again, can be generated. But then we know that heat is only a molecular vibration, a quivering motion of the atoms. Combustion and chemical action generally can set the atoms vibrating, and the body is then said to be hot. Vibrating atoms also set the ether quivering, and thus generate radiation or what we call light. These things are therefore not fundamental entities, but are peculiar states of motion affecting pre-existing things.

Electricity, however, does not appear to be of that kind, though it may be rash to assert definitely that it is not. All that we know is that we have not yet discovered any means of generating it—

neither by setting something in motion nor otherwise. A unit electric charge may turn out to be a peculiar local extremely-minute whirling motion of the ether, but that discovery has not yet been made. It seems to be a localised strain in the ether of some kind, something that is ready for, or liable to, locomotion. An electron is certainly the nucleus of an electric field, by which it is susceptible of easy control and extraordinarily rapid acceleration; but we know no way of making or coagulating such a nucleus out of ether. We know no more how to make an electric charge than how to make an atom. We encounter ready-made atoms, and deal with them in daily life; we encounter ready-made electric charges, and deal with them in the laboratory. Similarly in many substances we encounter ready-made molecular currents, which we are pretty sure are revolving electrons, and by making them face in one direction we can bring out and display the latent phenomenon of magnetism.

But since electrons are so docile, we can immediately go on to ask whether we have no means of setting them in motion, making them revolve in large orbits, and thus generating perceptible magnetism *de novo*. The answer is Yes. Means have been found of generating electric currents, and thus making electromagnets, which have the same properties as natural magnets, but are still more under control, and are not limited in strength.

That is true, now; but until the discovery of electromagnetism at the beginning of the nineteenth century it might have been said that we could not generate magnetism at all, that all we could do was to utilize existing magnetism, and transfer it from one piece of steel to another, or from the magnetic iron-ore in which it was originally discovered. In other words, we should have supposed that there was no magnetism without a pre-existing magnet; but we should have known that, given one magnet, we could produce any number of others, by stroking pieces of steel against it, the original magnet being no weaker than before, and that there was no limit to the number of magnets that could thus be made.

It is clear that statements can be made about Life, quite analogous to those which could be, and probably were, made about magnetism before the discoveries of the nineteenth century. No magnetism without antecedent magnetism; any amount of it producible; but no way of generating it *de novo*. Just so we know of no life except by the agency of pre-existing life: all attempts at spontaneous generation of life have failed. When life appears, no matter of how lowly a kind, as when food goes mouldy, or when putrefaction or fermentation sets in, it is always found that a germ of pre-existing life has entered and grown. If all such germs are excluded, no life appears; but given the influx of life, at all, any amount can make its appearance; there seems no limit: just as from a given magnet any number of magnets can be produced. So from a single plant there can be any number of seeds; and each seed, properly treated, would germinate and grow a plant; and so on without limit.

For the manifestation or appearance of life, suitable conditions must be supplied; but life is not explained by the conditions which it utilises for its epiphany. It is not really generated, it is embodied in matter, and is passed on from one organism to another. We cannot produce life *de novo*; though the discovery of how to produce magnetism *de novo* has been made. Galvani and Volta discovered how to produce an electric current. Such a current, sent round a coil of wire, magnetises a steel core without any obvious pre-existing magnetism. At present we have no clue to the generation of life; yet somehow life certainly has appeared on the earth, whereas at one time the earth was barren and at so high a temperature that no life could have existed. Life must have appeared *de novo* in the past, and (I repeat because it is curiously suggestive) what has happened in the past may possibly be happening in the present, and may be better understood and controlled in the future. We have now learnt that magnetism is due to electric charges in motion, and that there is no other magnetism—none save that accompanying an electric current; which, however, may be infra-microscopic in size. As soon as we know what life is, we may have some chance of entrapping it and experimentally displaying it in association with matter: at present we have no idea. So at present the doctrine of what Huxley called Bio-genesis holds the field; we know of no life except through the agency of antecedent life.

Life seems more analogous to magnetism than to electricity; for its amount varies, tends to increase, and can be made to increase by suitable arrangements. Like magnetism, life is illimitable in amount; whereas electric charges are fixed in amount, we know of no means of generating them; we only know how to set them moving. We have not yet learnt how an electron is constituted. Still less have we learnt of what life consists. To me it seems to be something pre-existing in the ether, which is able to enter into relation with matter when matter has attained a certain complexity; something which can endure that connexion for a time, and then depart whence it came. Some think that it is merely an appearance due to the complexity of chemical organisation; whereby the operations of assimilation, growth, and reproduction surprisingly follow. Others think that, though a real entity, life is something which cannot be explained in terms of physics and chemistry—that is, in terms of the constitution of matter—but that it involves something more, which at present we do not understand. And they are strengthened in that view by the familiar fact that life at a higher stage blossoms out into mind and consciousness, which are really the only things of which we are primarily aware, and which are clearly something quite different from any interaction of molecules, atoms, or electrons; for mind is able to deal with those things in an independent manner, to understand them, make experiments upon them, and generally to dominate the material world.

For the display of mind and consciousness here and now, a highly organised form of matter, such as exists in the brain, seems necessary. But I judge that mind controls and has produced that organisation, acting on matter from its vantage-ground or home in the ether; though I admit that we do not yet know whether mind and consciousness are related to the ether in the sort of way we have gradually found that everything else is related to the ether. We do not know for certain whether these ideal or psychic things need have any physical basis at all. The question is an open one, which can only be solved by further inquiry. Meanwhile we are among a group of questions which are dealt with in psychology and philosophy, and are outside the present scope of physics. He is wisest who keeps his mind open to these problems, and is ready to welcome any fresh facts and any channels of information, so as gradually to learn more about a region of future science which at present is only in its infancy.

Matter and Ether

Meanwhile, in the absence of knowledge, we must take refuge in speculation, being careful to remember that, though not baseless, it *is* speculation, and is therefore devoid of scientific authority. On those terms, fully admitted, I will give my present view about some of these high matters, though the details which have led me to them must be treated elsewhere. But first an orthodox preamble.

First of all, then, I discriminate, as I think most of us do, between the adjectives "physical" and "material"; giving a narrow interpretation, for the moment, to the phrase "the material world" as meaning the world of matter. The universe, even on its physical side, does not consist of matter alone. Matter is all that appeals to our senses; but we know that in addition to matter and its various affections, such as force, momentum, elasticity, density, weight, heat, sound, and the rest—all of which have been known from time immemorial—there are other things which are not matter, such as light and electricity, which, though in a sense known for a long time, have been only begun to be studied and analysed and partially understood during the last century. These things interact with matter, and whether they be called material or not is partly a question of nomenclature : they are certainly physical; they have to do not primarily with matter but with ether. To them I would add gravitation and cohesion, which, like light, are concerned with or are located in the interspaces between particles of matter; that is to say, gravitation and cohesion and elasticity are due to properties of what we call vacuum, or what some call space, and others prefer to call ether, because directly space is endowed with physical properties it becomes something more than the mere geometrical idea of extension or room to move in. Ether it is that holds the masses of matter together, welding the planetary masses into a cosmos by gravitation, and the molecules into solids and liquids by cohesion. Ether, moreover, is susceptible of a periodicity analogous to vibration, by means of which information is received about the constitution and movements of otherwise hopelesssly inaccessible and distant objects. Ether is responsible for all potential energy, and is continually interchanging energy with matter; in association with which it takes the kinetic form.

I might add to the other immaterial things

the idea of time or duration; an idea essential to evolution, and indeed to our whole notion of permanent existence. But if we touch on "time" we are getting beyond our present scope. I only mention it as an instance of innumerable other immaterial things which we are conscious of, though so far as I know we make no attempt, or at least no successful attempt, to disentangle their ultimate nature. To suggest what I mean by "innumerable other things", I merely catalogue the words, Beauty, Aspiration, Intelligence, Faith, Affection, Will: the significance of these terms being such as to clamour at the absurdity of the idea that the universe consists of matter alone, or that it is ultimately explicable and intelligible in terms of physical science alone, even when expanded and interpreted in its widest sense. All that at present I want to insist upon is that matter does not exhaust even the physical universe, and that the ether, or whatever is equivalent to it, must be taken into account.

But, again, all these ultra-material things are only known to us in their association with matter. We have inferred or discovered that light is an etheric vibration, but we do not see or sense the vibrations: what we *see* are the material objects upon which light falls. And we do not even see them until their re-emitted, scattered, and modified light has encountered the nerve terminals of the retina; just as we do not hear a distant bell till the vibrations have stimulated the nerves of the ear. We are thus dependent for our information about the external world on our material organism, which indeed is part of that world. But the amount of information which we can *infer*, from simple sense indications, is prodigious, and out of all proportion to the stimulus which evokes those inferences. The inferences, I must repeat once more, are affairs of the mind. They belong to a different order of things, and are not the result of sense indications; else the animals would know as much about the Universe as we do.

Ether and Life

Now I come to notions of a less common-sense character, in the hope that posterity, if it ever reads them, may perhaps be either interested or amused.

I postulate, then, as the one all-embracing reality on the physical side, the Ether of Space. And I conceive that in terms of that fundamental physical entity everything else in the material universe will have to be explained. To me the ether is a continuous substance, far more substantial than any matter. It fills all space; though we have no real knowledge of its constitution, for it is too fundamental to have its constitution expressible in terms of anything else. It seems to be analogous to a perfect incompressible fluid in a violent state of minute circulatory motion, imperfectly conceivable as ultramicroscopic vortices circulating or spinning with the velocity of light. For Lord Kelvin and G. F. Fitz-Gerald and others have tried, with partial success, to show mathematically that such a rotational fluid would transmit transverse waves at a speed comparable or even identical with its intrinsic speed of circulation. This velocity, which has been accurately measured, and is commonly known as *c*, appears to be a fundamental constant in nature. According to one theory the speed of etheric circulation bears the same relation to the speed of wave-transmission as a length equal to three sides of a square bears to the diagonal of the same square. (*Phil. Mag.*, October 1887, p. 350.)

A velocity necessarily implies or suggests the idea of time. Speed is in fact what we directly apprehend in any ordinary motion; space and time are abstractions derived from locomotion and speed. To me space and time seem likely to be also physical realities, but certainly they are mental abstractions, based on experience and generalised from it, like the ideas of force and energy.

If the ether is constituted as I suggest, it must be the seat of enormous energy; not necessarily infinite, but far beyond any amount of which we have any conception. All the energies that we experience in matter are but a minute or residual fraction of the etherial energy of which they are a feeble manifestation.

My speculation is that this boundless ether, thus full of energy, is utilised and is impregnated throughout with something that may be called Life and Mind *in excelsis*, that it is the home of the ideal and the supernal, and that all the life and mind we are conscious of is but an infinitesimal or residual fraction of this majestic reality. I conceive of the ether as the vehicle or physical instrument or concomitant of Supreme Mind. It may be that "Spirit" is the better term, that Spirit permeates and infuses everything, and that it controls, sustains, and is commingled with the visible and tangible frame of things.

So said Virgil long ago, in the Sixth Book of the *Æneid*. And, coming to extremely recent times, I find in a recent issue of the *Radio Times* a striking anti-war article by Sir Ian Hamilton, in which he speaks—in connexion with the Cenotaph celebrations of the British Legion on Whit-Sunday—of "that mystical instinct which tells us that the universe not only supports life, but is itself pulsating with life". What exactly was in the mind of the distinguished soldier when he wrote this I do not know, but in myself a conviction has gradually formed that the physical ether, itself co-extensive with the universe, is literally and physically squirming or pulsating with life and mind. As if we might regard it as a great reservoir of Life whence individualised fragments can from time to time be drawn, as from a store of raw material: if for the moment, and from the merely human point of view, it is permissible to speak of the fundamental substance in such terms as raw material, even though at the same time we recognise it as likely to be the physical vehicle of the Supreme.

For the Universal Supreme Spirit must have many aspects, many manifestations, perhaps even many modifications. Here and there small portions of it have become individualised; in some analogy with, and perhaps by aid of, the individualised electronic modifications of the universal ether; so that throughout space—and not only on lumps of matter—there are individual psychic entities of various grades, down even to those which are akin to our own standard, and even lower than that, down to their rudiments in the amœba.

Little can I say about the individualised portions of the Universal Spirit, beyond what everybody knows by his own experience; but concerning the individualised physical entities in the universal ether, I can conjecture something more. I not only conjecture, I feel as if I knew, that portions of the ether are so to speak individualised, by the whirling motion taking on here and there an identity or individual form, which thereafter is susceptible of locomotion, and can be recognised or identified as it moves from place to place. The circulatory speed being so modified and slowed down in its passage as to give the effect of locomotion; a term which in the unmodified or unidentified ether would be of no significance—an unmeaning term—since "place" is hardly predicable of an infinite continuum without identifying singular points.

To go briefly into further detail. The singular points or locomotive particles are of two kinds, one kind as I have thought hollow, the production of the hollow being itself a sign of great energy, so that an extra strain is set up surrounding the hollow—a strain region which we have become familiar with as an electric field—the seat of a special kind of potential energy. The energy of this field endows the hollow at its centre with what we call inertia; and as it moves from place to place its energy travels with it, being localised chiefly in the immediate neighbourhood of the hollow, which we have now learnt to call "an electron". An electron is a thing with special electrical properties, and forms an identifiable particle, which may agglomerate with others, and thereby is responsible for nearly all that we know, not only in electricity, but in chemistry too.

The other singular point, or locomotive centre,

which is necessarily produced at the same time as the hollow, appears to be extra solid and substantial, and is also surrounded by an electric field. This field represents the strain of its production, and enables it to be localised and identified; but it possesses a nucleus of such surprising solidity (whatever that may mean) as to have an inertia far greater than that of an electron—greater as a hundredweight is greater than an ounce—an inertia of measurable amount, but of unknown cause. The cause of the large mass might be that the nucleus is extra small and concentrated; or its inertia may be so great because it is spinning at a prodigious rate; or (fancifully) because substance removed from the interior of the electron is crammed into the proton; or the extra massiveness may be due to something as yet unguessed. Anyhow, this massive particle, which is electrically equivalent to the other, though endowed with greater total energy, seems to constitute the primary unit of matter, and has recently been named "proton". About it there is much more to be discovered. What we know is that every proton attracts every electron, and that they thus tend to combine into atomic systems, akin to the systems which we have long been familiar with, on a gigantic scale, as the solar systems and other units of astronomy.

Compared with the vastness of space the volume occupied by atomic or molecular systems is small; while the space occupied by the extremely minute positive and negative particles themselves is utterly insignificant. Nevertheless they have their function to perform in space, they agglomerate together in immense numbers, forming diffuse visible masses. "Visible", because their sudden motions and collisions—their internal convulsions—disturb the rest of the ether in which they exist, and throw it into vibration. When these vibrations fall upon our eyes, through the instruments by which we aid our vision, we infer the existence of whirling clouds of matter. We call these gigantic masses spiral nebulæ, or other conventional names, and perceive that some of them are at incredible distances, so that their light may have taken millions of years or even millions of centuries to come.

The rest of the evolving process is too long to go into in detail. Suffice it to say that it has been worked out by astronomers, notably by the great cosmologists, like Sir James Jeans of our own country, supplemented and confirmed by some in America; and we thus learn about the birth of stellar systems, and ultimately of isolated suns and stars as we see them. That is to say, the process is now being studied by which, out of modified ether, have been formed those bright globes which have always attracted our attention and interest; more especially so now that at length those other flaming more distant masses have come within the indirect purview of human senses and consequent ratiocination.

But, going back to the early stages of stellar formation, not yet has any vitalising power from the ether entered into relation with the modified portions: not yet has life been able to associate itself with these flaming masses; they are not in a condition fit for the habitation of life. Here and there, however, a surprising event has occurred; a sun has thrown off a streak of matter, which has consolidated into what we call planets. These, being small bodies, cool down with some rapidity, and ultimately reach a temperature only a few hundred degrees above absolute zero—a temperature within the narrow range at which water can exist in the liquid state. The planetary masses (or at least one of them) are kept at this reasonable temperature by radiation from the central hot body which emitted them. Molecules are now able to form and to grow in complexity; chemical compounds, and ultimately protoplasm, make their appearance.

Now occurs an incredible thing, or only credible because it must have actually happened. The localised and identifiable groups of matter-particles become able to receive and incorporate some of the previously unidentified life and mind of which the general ether is full. Life thus associated with individual particles becomes itself an individual, isolated from and so to speak forgetting its previous unidentified existence. Somehow by its own inherent properties it makes use of the matter and energy which it finds to hand, and builds up a specific organism; at first of a lowly kind; afterwards, through a long process of evolution, into what we have become familiar with as the vast variety of plants and animals.

Gradually, and through long stages, some of the life develops into mind; or, rather, mind itself becomes individualised and incarnate in the most highly developed of the organisms; and thus begins the reign of individual consciousness. The development of sciences which seek to understand the whole process, so far as it has come within their ken, now also begins. Astronomy works out the laws of all the masses visible at a distance. Botany studies the plants, Zoology the animals, Biology takes a comprehensive view of life in general, while Psychology seeks to interpret the element of mind. Physics and Chemistry deal with the inorganic substances in general; while Geography deals with the particular mass of matter which, by reason of its low temperature, humidity, atmospheric and other advantages, has been favoured by gradually becoming the habitat of conscious beings who can admire the display and speculate on the meaning of existence.

That is where we are now: we little know what is ahead, or to what it is all leading. And then some presumptuously try to say that they understand the general character of the whole process, and that it is a self-acting one, due to the agglomeration and complexity of matter. They may go on to say that the ether does not exist, and that the idea of any life or mind existing or persisting out of association with a material organism is an absurdity. They do not see that the really strange problem is how life and mind came into association with matter at all; they will not entertain the notion that they themselves are incarnations of a persistent immaterial entity for a brief period; and they often deny the possibility of any mode of existence other than that to which they are accustomed.

The fact is that mere survival or continuity of existence, when regarded from the proper point of view, must be admitted as inevitable; the only rational question is about individual survival. And that question must be answered by an investigation and scrutiny of facts which are gradually forcing themselves more and more on our attention, but which for the most part are not yet studied by any of the orthodox sciences. They are in fact too simple, and are ignored by "the wise and prudent".

My own conclusion, based on those same elementary facts, is that Personality—when it exists—including Character and Memory, is certainly a persistent reality. I reach that conclusion not by argument, but by direct personal experience, as simple as are the experiences of daily life. It is an elementary deduction from experience, and needs no elaborate reasoning or argumentation; though I do now seek to justify and rationalise this elementary deduction by physical discussion and analogies, in the hope of gradually illuminating the nature of life and mind in general, and of some day understanding the method of their interaction with the physical and material world.

As a preliminary, let us see whether we can gain some idea about what is now known concerning the nature of matter.

CHAPTER III

THE NATURE OF MATTER AND RADIATION

To my mind the greatest scientific event during the present century is the establishment of the Electrical Theory of Matter and the recognition of matter as one of the forms of energy. Up to this century matter had always been thought of as a thing *sui generis*, unresolvable into anything simpler, the amount of which was absolutely conserved and unalterable, being never either generated or destroyed. But with the discovery of the electron in the closing years of the nineteenth century, and the notable initiation of the idea that the atom of matter was composed of electrons and protons—that is, of negative and positive electric charges, and nothing else—the way was open, first, for the perception that matter was an electrical or, in other words, an etherial phenomenon; and, next, for the perhaps remote possibility that by the clashing together of positive and negative charges the atom of matter as such might disappear and result in the production of some other form of etheric energy, primarily in the form of radiation. In so far as mass or inertia could be explained electrically, it could not be constant, but must be a function of speed; and this was a revolutionary supplement to the mechanics of Galileo and Newton.

One outcome of the Theory of Relativity is that there is no absolute distinction to be drawn between matter and other forms of energy, and that what we perceive as motion of matter is but a residual etheric phenomenon in which some fraction of the unmodified ether is entangled with the modified portion that we call matter, so as to increase its mass and give us the effect of movement.

Internal Movement

It is very difficult to explain this in any intelligible manner except by rough analogies. It has long been known that in any fluid or gas the atoms were moving rapidly with a speed comparable to that of bullets, and that what we call a wind or a current is but a slight trend of the motion of the particles in one direction; as when a swarm of gnats, flying vigorously among themselves, slowly drifts along in one direction. Anyone looking at the swarm from a distance, and not seeing the individual insects, would not know about their internal motion, he would only see the locomotion of the swarm, which is really only an insignificant and subsidiary trend in one direction. So it is with a wind or current. The speed of even a hurricane is but a small fraction of the motion of the constituent particles.

Matter and Ether

Applying this analogy to the ether, it is becoming plausible and even probable that the ether is circulating internally, in a fine-grained and quite imperceptible manner, with a prodigious velocity, called the velocity of light, and that it is by some peculiarity or modification of this motion that the electrons and protons are constituted, and that therefore the atoms of matter exist. A very slight trend of this circulation in one direction would then constitute what we perceive in the gross as the locomotion of matter, as of a railway train or an aeroplane or the still quicker motion of a planet. These perceptible locomotions, so humanly important, are trivialities largely dependent on the circumstances of an observer, and are not easy to specify absolutely. Nothing that varies for different observers can be fundamental. The probability is that we shall ultimately find that the material universe is constituted of ether in various stages or varieties of circulatory movement, and that this velocity—measured as the velocity of light—is the only absolute velocity, the same for all observers; though, as a matter of fact, the etheric circulation is to all observers at present entirely imperceptible. It has to be inferred, and no experiment that has yet been suggested can verify it.

Nevertheless space, instead of being a mere background for events, is now perceived to be the scene of vast activity, a storehouse of energy; of which energy, matter, hitherto so absorbingly attended to, is but a small fraction, so that it is beginning to recede and become itself a background for the display of radiation or etheric energy in its various and manifold forms.

Clues of the Stars

How, then, can we say that matter is one of the forms of energy, and that the disappearance of matter would result in radiation? We say this on the strength of our observations of the stars, which are brilliantly luminous bodies of great size and immensely high temperature, in the interior of which strange processes are occurring. Matter is there both disappearing and being built up, and the radiation or light that we see is but an escaping portion of the great turmoil of energy within. The interior of a star is so violent and energetic that it gives away its secrets, and enables us indirectly to perceive what is going on, more clearly and substantially than we can perceive what is going on in an apple, or even in a pebble. Ordinary matter, as we know it, is quiet and restrained, and keeps its secrets; its substance is conserved and apparently perpetual; but the behaviour of matter in the interior of a star is quite different from that. It generates a blaze of energy, emitting it into the ether in all directions; and by analysing that radiated energy the great astronomers of the present day are beginning to feel that they know more about the interior of a star than they do about the intimate construction of their own bodies or about the interior of the earth. They are thus able to begin to disentangle or unravel the ancient problems of the nature of matter, the nature of energy, and, incidentally, the general structure of the material universe. Only Life and Mind still remain outside their ken.

Ultra-Modern Views

Changes are so rapid in modern Physics that every mode of expression is liable to become antiquated in a few months. Electrons, for instance, used to be thought of as particles of definite size moving through the ether; but now what is called wave-dynamics has begun to monopolise the field, and a thing which seemed like a particle can be equally well regarded as a wave; not perhaps an ordinary wave, but something more like a group wave, such as will be mentioned later. The quantum theory has effected the converse representation, for ether waves are now perceived to have in many respects the properties of particles. Their energy appears concentrated, in an individual manner, so that it all acts together if it acts at all. The result is that when a wave encounters an obstacle, it acts like a projectile, with a hit or miss kind of result, equivalent to the absorption or emission of all or none. Hence the ancient distinction between particles and waves, or more generally the distinction between matter and radiation, tends to be instructively and intelligently confused.

Some Curious Notions Connected with Wave Mechanics

As an illustration of the way in which what had hitherto been considered particles can now be treated as waves, and vice versa, I am going to refer to a couple of papers which have appeared in the most recent Part of the *Proceedings of the Royal Society*, issued in December 1929.

The first is by N. F. Mott, of St. John's College, Cambridge, assisted by Professor C. G. Darwin, F.R.S., and its title is "The Wave Mechanics of Alpha-Ray Tracks". The problem it attacks is, How can the luminous track in a Wilson Ray machine—which is known to be made visible by the condensation of saturated aqueous vapour on ions liberated from the atoms ionised during the flight of an alpha-ray particle shot off by radium—how can that track be straight, if the alpha-ray projectile is a wave instead of a particle? The problem strikes one as rather akin to the quite-differently-solved century-old problem of how on the wave theory light could travel in straight lines, so as to cast sharp shadows. I cannot suppose that the solution of the alpha-ray problem is as yet complete; but the argument depends on a calculation of the probability of a second encounter, between a portion of a wave-front and another atom in its path. I do not follow the reasoning, but the conclusion seems to be that the chances of such a second ionisation are remote or nil, unless the second particle encountered lies in a straight line with the first ionised particle; that is to say, in a straight line produced, from the radium source, through the first particle that happens to be ionised in any given direction. In other words it is attempted to be shown that any number of atoms can be ionised by an alpha-wavicle if they all lie along a straight line radiating from the source, so as each to be as it were in the shadow of the one before it; and thus the rectilinear Wilson tracks of the alpha-particles are supposed to be accounted for on the wave theory, without postulating any actual mechanical projectiles.

The other paper which has attracted my attention, in the same issue of the *Proceedings of the Royal Society*, is by Professor Mosharrafa of Cairo; it is communicated by Research-Professor O. W. Richardson, F.R.S., and is called

"Wave Mechanics and the Dual Aspect of Matter and Radiation". A suggestion is here made, on the basis that all phenomena travelling past us with the velocity of light are called by us "radiation", while the localised events which travel much more slowly, or not at all, are called by us "matter"; though they may all be supposed to be fundamentally, and equally, of the nature of etheric waves. The question therefore is raised, How would things be regarded by an observer who is himself travelling with the velocity of light? And it is suggested that, in his case, that which we call radiation, which might [doubtfully] be either accompanying him or sauntering past him, would appeal to him as "matter"; while the things that we call matter, which would [certainly] be flying past him with the velocity of light, he would stigmatise as "radiation".

The idea is an ingenious one, worthy to be treated by a Dean Swift; and yet, though a speculation, it is by no means an absurd speculation. It is the result of a portentous piece of abstract mathematical reasoning, such as is generally supposed to justify a considerable and sometimes surprising amount of speculative deduction. Abstract wave-mechanics, applied to an electron, seems to lead to the conclusion that the surface of an electron is coincident with a Schrödinger wave surface.

Whatever we may think of this semi-identification between particles and waves, there is certainly plenty more to be discovered about the inter-relation between matter and radiation; and a discussion of this and other cognate matters—all emphasising the importance of the ether—must lead us into a region not disparate from but rather beyond ordinary physics. Of which region we must now attempt a survey.

CHAPTER IV

A SURVEY AND A CRITICISM OF MODERN PHYSICS

WITH AN ATTEMPT TO FIND A PERMANENT PHYSICAL BASIS FOR LIFE AND MIND[1]

Addressed to Philosophers

HAD I been asked any time last century to discourse to philosophers on a subject entitled "Beyond Physics", I should have taken it as meaning "Metaphysics", and have declined on the ground that they knew more about it than I did. But now that the subject has been suggested to me in the second quarter of the twentieth century, I realise that a certain number of physicists, especially mathematical ones, have allowed themselves latitude in a semi-philosophical direction, have thrown our old physical ideas into some confusion, and have devised a number of complicated abstractions which they are able to deal with by recondite methods invented for other purposes by the pure mathematician. Of these they do not attempt to form any mental image, and do not encourage us to try to do so. They mistrust the method of physical imagery, or what were called "models", which constituted

[1] An expansion of an Address given to the British Institute of Philosophical Studies in July 1929, under the requested title "Beyond Physics".

one of the main methods of investigation in the nineteenth century; they are content to deal with abstract symbols, which stand for things largely unknown. The symbols may represent something real, but they mistrust the term "reality", and decline to define it. The younger modern physicists are content to work in a nebulous region of unknown and perhaps unknowable entities, remote from our ordinary conceptions and everyday experiences; so that, whereas the old idea of explanation consisted in trying to express mysterious or recondite phenomena in terms of something with which we were more familiar, they now point out that our apparent understanding of those simpler things was more or less of an illusion, that they furnished no valid explanation at all, and that the physical and engineering conceptions or habits of thought on which we depended were largely misleading. They claim that the things we observe and measure would be observed and measured differently by different observers, and that no one is able to say that one observation is better than another. The measurements made by observers with unknown motions are all equally near, and perhaps equally far from, the truth: they are all "relative" except in a few salient and unusual instances.

The Bearing of Modern Science on an Idealistic Interpretation of the Universe

I find that the term "Beyond Physics" was actually used by one of the most eminent of these new explorers, for it occurs on page 344 of the published (1928) version of Eddington's Gifford Lectures, *On the Nature of the Physical World*, delivered in Edinburgh in 1927. I quote part of a passage:

> "The physicist now regards his own external world in a way which I can only describe as more mystical, though not less exact and practical, than that which prevailed some years ago, when it was taken for granted that nothing could be true unless an engineer could make a model of it. There was a time when the whole combination of self and environment which makes up experience seemed likely to pass under the dominion of a physics much more iron-bound than it is now. That overweening phase, when it was almost necessary to ask the permission of physics to call one's soul one's own, is past. The change gives rise to thoughts which ought to be developed. Even if we cannot attain to much clarity of constructive thought, we can discern that certain assumptions, expectations, or fears are no longer applicable.
>
> "Is it merely a well-meaning kind of nonsense for a physicist to affirm this necessity for an outlook beyond physics? It is worse nonsense to deny it."

I do not continue to quote the witty remarks with which this paragraph is followed up, though I would call attention to his next page, where he puts a dilemma before those who would explain everything, including even nonsense, by processes in the brain, and asks what criterion they have on that basis for discriminating between a right and a wrong brain-process, or between "ought" and "ought not"; what criterion they have for nonsense, on their principles, and by what right they would condemn it. As, for instance, why should we object to a wrong arithmetical result produced by an ill-working brain; why desire a correct answer?

> "A physical machine cannot esteem or want anything; whatever is fed into it it will chaw up according to the laws of its physical machinery. That which in the physical world shadows the nonsense in the mind affords no ground for its condemnation. In a world of æther and electrons we might perhaps encounter *nonsense*; we could not encounter *damned nonsense.*"

In reading Eddington and some other of the more mystical or abstract physicists who live in a world of unknown and possibly unknowable entities, I am sometimes tempted to ask whether we are in any way making a return to the discarded philosophy of Herbert Spencer. There was a time, more than half a century ago, when I studied philosophy to some extent under Croom Robertson at University College, London, with the textbooks of Bain on the Sensations, the Emotions, and the Will, that I became enamoured of Herbert Spencer's *First Principles.* I remember copying out large portions of that book, so as to impress it on my mind. But gradually I found that Herbert Spencer's physics were shaky, and occasionally what we called "wrong", and I ventured to regard his contention for an unknowable substratum as an unnecessary closing of the door. But perhaps this perception of an "unknowable" was only premature; perhaps the present obliteration of the distinction between right and wrong measurements is a step in the same direction as the philosophy of Nietzsche, who attempted to go "beyond good and evil". I am not competent to judge how far Spencer may turn out to have been justified in some of his contentions by the advance of physics, but it may be worthy of consideration by professed philosophers.

I must hasten to add that Eddington in his own philosophy does not succumb to the idea that the physical world contains everything of importance, and that success in attaining results is the only test of truth. He himself is ready to condemn nonsense on grounds beyond physics; he is by no means willing to limit the universe to chemical and physical processes, however successful they may be: and so he adds:

> "My own concern lest I should have been talking nonsense ends in persuading me that I have to reckon with something that could not possibly be found in the physical world."

He thus admits and defends a charge which may be brought against his lectures, of admitting some degree of supernaturalism, and says that "in so far as supernaturalism is associated with the denial of strict causality, I can only answer that that is what the modern scientific development of the quantum theory brings us to". Consciousness, after all, is part of nature; and though in physics consciousness may be regarded as "something which unfortunately has to be admitted but which it is scarcely polite to mention", Eddington evades the difficulty by limitation of area; preferring to restrict physics to a cycle of operations going round and round in a cycle, complete in itself, without the more immediately apprehensible human faculties entering in. He claims to have associated consciousness with a background untouched in the physical survey of the world, and thus incidentally to have given the physicist "a domain where he can go round and round in cycles without ever encountering anything to bring a blush to his cheek".

> "It has been . . . the purpose of our discussion to secure such a realm where scientific method may work unhindered. . . . The accusation is

often made that, by its neglect of aspects of human experience evident to a wider culture, physical science has been overtaken by a kind of madness leading it sadly astray. It is part of our contention that there exists a wide field of research for which the methods of physics suffice, into which the introduction of these other aspects would be entirely mischievous."

I quote these inadequate samples, in fairness to Professor Eddington and his co-thinkers, as showing that they do not regard their mathematical abstractions as a valid ground for the philosopher and the theologian to base their theories upon. He considers that such a basis would be unprofitable and even dangerous, inasmuch as it would leave enduring entities at the mercy of scientific revolution and changes of thought. He partitions off the domain of physical science,—limiting it, as I think unduly, to its metrical aspect,—claiming that the data which mathematical physicists employ are based solely on laboratory experience and the readings of instruments. He would limit the domain of physics proper to "pointer-readings". This, he says, is the quantitative aspect with which mathematics deals, and it necessarily excludes great tracts of direct human experience which are not metrical at all; but he is careful to explain that in excluding them from the limited scientific domain he is not making the mistake of denying their existence and high value. Beyond our physical and metrical apprehensions he would admit our æsthetic and religious convictions and intuitions as equally real, equally valid, and still more comprehensive: not a mere colourless deduction from or extension of our physical process, but something approached from a different side altogether. If we approach the spiritual world from the side of the material, all we can assert of it would be insufficient to justify even the palest brand of theology:

> "But the spiritual world as understood in any serious religion is by no means a colourless domain . . . the attribution of religious colour to the domain must rest on inner conviction."

And here is a quotation which I imagine would be heartily approved by your President, Lord Balfour; for as I understand his philosophy, he has urged something of much the same kind, at many times and in many places:

> "We should not deny validity to certain inner convictions, which seem parallel with the unreasoning trust in reason which is at the basis of mathematics, with an innate sense of the fitness of things which is at the basis of the science of the physical world, and with an irresistible sense of incongruity which is at the basis of the justification of humour. Or perhaps it is not so much a question of asserting the validity of these convictions as of recognising their function as an essential part of

> our nature. We do not defend the validity of seeing beauty in a natural landscape; we accept with gratitude the fact that we are so endowed as to see it that way."

Otherwise, if we were to base our religious convictions and our sense of free will on the discoveries of modern science and on the progress recently made towards the indeterminateness of certain atomic processes, and other curious developments of the latest speculations in physics, we might be led to the preposterous conclusion that "religion first became possible for a reasonable scientific man about the year 1927".

I emphasise this caution because I myself am tempted to go further than Eddington, and to see in many of the conclusions, now either drawn or in process of being drawn by modern science, some justification and some assistance towards a possible comprehension of the universe in its wider and larger aspects. I myself have been led in these directions by psychical and other investigations, which form no part of Eddington's scheme, and which at present lie outside the domain of physics. My aim is to incorporate them with physics, and to seek to show a connexion between physics widely interpreted and the region beyond physics. The task is a difficult one, and may be impossible; but then, failure need throw no doubt upon the validity and genuine consistency of the inferences drawn in different departments of human thought. A failure to unify them is a human failure, not necessarily embedded in the nature of things at all.

Nevertheless, in spite of his unwillingness to base anything really serious upon the outcome of physical discovery, instead of upon direct apprehension, it is clear that Eddington does regard with some hope, and even with some complacency, the idea that modern physics has opened a fresh avenue towards Indeterminism and Free Will, by its apparent overthrow of the strict causality in which hitherto it had been bound. Those pointer-readings, on the limitations of which he insists, may after all be more than mere readings of a scale; they may be unwittingly pointing, like a finger-post, in some philosophic direction. For he speaks as follows:

> "If our expectation should prove well founded, —that 1927 has seen the final overthrow of strict causality by Heisenberg, Bohr, Born, and others—the year will certainly rank as one of the greatest epochs in the development of scientific philosophy. But seeing that before this enlightened era men managed to persuade themselves that they had to mould their own material future, notwithstanding the yoke of strict causality, they might well use the same *modus vivendi* in religion."

His method of averting a conflict between science and religion is for both sides to confine themselves to their proper domain. To take one of his own examples: Most people think that there is a future non-material existence in store for us; this, which may be called Heaven, used to be thought of as somewhere in space: now it is regarded as somewhere in time.

> "(All the meaning of the belief is bound up with the word *future*; there is no comfort in an assurance of bliss in some *former* state of existence.) On the other hand, the scientist declares that time and space are a single continuum; and the modern idea of a Heaven in time but not in space is in this respect more at variance with science than the pre-Copernican idea of a Heaven above our heads. The question I am putting is not whether the theologian or the scientist is right, but which is trespassing on the domain of the other? . . . Science and theology can make what mistakes they please provided that they make them *in their own territory*; they cannot quarrel if they keep to their own realms. . . . It is not so much the particular form that scientific theories have now taken . . . as the movement of thought behind them that concerns the philosopher. Our eyes once opened, we may pass on to a yet newer outlook on the world, but we can never go back to the old outlook."

I have extracted these passages, and would like to extract many others, from Eddington's brilliant and accessible book called *The Nature of the Physical World*, because I want to avoid even the appearance of seeming to claim his great authority in support of any speculations of my own. I sometimes agree with him, and sometimes disagree. He is dealing in a vivid and illuminating manner with very difficult subjects. The main content of the book is a treatise in ordinary language on the most advanced and most modern physics. In reading him you will find that he makes strange and sometimes incredible assertions. He purposely puts things in a way that will attract attention, and may lead people to think sometimes that he cannot know what he is talking about. Some of the assertions will sound like nonsense. But as another great mathematical physicist (Prof. Whitehead) has said: "Heaven only knows what seeming nonsense may not to-morrow be demonstrated truth."

What I want to guard readers against is any idea that Eddington is talking careless nonsense. His utterances represent the outcome of profound knowledge, such as is possessed by few other men; and to criticise or disagree with him is no light matter. There is a mass of unexpressed meaning behind all his sentences. His paradoxes have a wealth of meaning behind them. I sometimes wish that he had illustrated his book by interleaved pages, which might be skipped, giving some indication of the mathematics on which his assertions are based. Let it be understood, however, that, though this would be a heavy task, it could be done; the mathematics can be found in other writings, either by himself or by others (*Space, Time, and Gravitation*, for instance); and students of the future will have to take all these writings into account. Even those which are being superseded are instructive, for they show the steps by which the further deductions have been reached. The whole science is in an active growing condition, though the intermediate stages may be ephemeral, like the tentative placing together of the dissected fragments of a picture in a jigsaw puzzle:

> "These revolutions of thought as to the final picture do not cause the scientist to lose faith in his handiwork, for he is aware that the completed portion is growing steadily. Those who look over his shoulder and use the present partially developed picture for purposes outside science do so at their own risk."

And in another place he says something similar that I will quote directly, for it may be taken as an anticipatory warning, to me and my like, not to attempt to extract psychical nutriment from tentative and uncertain stages in modern physics. Which nevertheless I am going to do.

Here is the quotation:

> "Schrödinger's theory is now enjoying the full tide of popularity, partly because of intrinsic merit, but also, I suspect, partly because it

is the only one of the three that is simple enough to be misunderstood.

. . . It would probably be wiser to nail up over the door of the new quantum theory a notice, 'Structural alterations in progress. No admittance except on business', and particularly to warn the door-keeper to keep out prying philosophers. . . . I do not see the least likelihood that his ideas will survive long in their present form."

Well, that is a weighty warning, and if my readers proceed any further with my exposition and speculation they will doubtless bear it in mind.

After all, neither Eddington nor anyone can remain uninfluenced by the far-reaching nature of modern discoveries. He says in his Preface that his philosophical views

> "can only claim attention in so far as they are the direct outcome of a study and apprehension of modern scientific work. . . . The idealistic tinge in my conception of the physical world arose out of mathematical researches on the relativity theory. In so far as I had any earlier philosophical views, they were of an entirely different complexion."

For my own part I could say much the same in respect of my psychical researches. But these, I must hasten to add, are entirely outside Eddington's scope; a fact which to me makes his conclusions all the more valuable, because when people arrive at similar results by different paths, it is an indication that they are proceeding on the whole in a right direction. Eddington does not base his spiritual outlook on the kind of facts with which I have concerned myself. He says on page 340:

> "A point that must be insisted on is that religion, or contact with spiritual power, if it has any general importance at all, must be a commonplace matter of ordinary life, and it should be treated as such in any discussion. I hope that you have not interpreted my references to mysticism as referring to abnormal experiences and revelations. I am not qualified to discuss what evidential value (if any) may be attached to the stranger forms of experience and insight. But, in any case, to suppose that mystical religion is mainly concerned with these, is like supposing that Einstein's theory is mainly concerned with the perihelion of Mercury and a few other exceptional observations."

I accept the analogy. In so far as psychic experiences are exceptional and peculiar, they serve as tests wherewith to confront ideas otherwise arrived at. They can be treated as of the nature of verifications; just as the advance of the perihelion of Mercury, and the deflection of a beam of light by a gravitational mass, were treated as verifications of Einstein's theory of relativity. If the deduction of the reality of a spiritual world is made on any grounds whatever, then that reality ought to have consequences which could be verified. And I claim that immediate contact or intercourse with the denizens of such a spiritual world is of the nature of verification.

New Outlook on the Material World

The idealistic world of the broadminded physicists—such as it is—is an indefinite vagueness deduced in an elaborate way from the manifest incompleteness of any other mode of regarding the universe as a whole. The material aspect of things is insufficient, and there are many indications of something beyond. The physicist has begun to go beyond physics into he knows not what.

> "No familiar conceptions can be woven round the electron; it belongs to the waiting list. . . . Descriptions of the phenomena of atomic physics have an extraordinary vividness. We see the atoms with their girdles of circulating electrons darting hither and thither, colliding and rebounding. Free electrons torn from the girdles hurry away a hundred times faster. curving sharply round the atoms with side-slips and hairbreadth escapes. The truants are caught and attached to the girdles, and the escaping energy shakes the ether into vibration. X-rays impinge on the atoms and toss the electrons into higher orbits. We see

> these electrons falling back again, sometimes by steps, sometimes with a rush, caught in a cul-de-sac of metastability, hesitating before 'forbidden passages'. Behind it all, the quantum *h* regulates each change with mathematical precision. This is the sort of picture that appeals to our understanding—no insubstantial pageant to fade like a dream."

So says Eddington with vivid imagery.

> "Nevertheless," he goes on, "the description of the processes must be taken with a grain of salt. The tossing up of the electron is a conventional way of depicting a particular change of state of the atoms which cannot really be associated with movements in space as macroscopically conceived. *Something unknown is doing we don't know what*—that is what our theory amounts to."

Well, it amounts to a good deal more than that; but still even in physics the essential nature of the units is unknown, and still more unknown must be the nature of a non-physical world. And yet some such world has been inferred. Here is where psychic experience enters. If the world exists, if we can get into contact with it, we may gain primary apprehension of it, and thus gradually our experience can become more definite. There are other avenues to truth than the metrical methods of physics. Eddington fully admits that there are avenues other than those of metrical science—though he would consider them non-scientific—but I want to go further and claim that these other avenues, or some of them, are open to scientific exploration. Biological methods, for instance, are often not metrical, but they constitute a part of science. So do psychological methods. Science is not limited to its metrical aspect. And in the exploration of reality every method must be used and every avenue explored. We belong both to the physical and to the unphysical world, here and now; we cannot afford to separate them and keep them in watertight compartments.

> "We are bound to claim for human nature that, either of itself or as inspired by a power beyond, it is capable of making legitimate judgments of significance. Otherwise we cannot even reach a physical world. . . . To make a start we must be aware of something . . . and we must be convinced of the significance of that awareness. . . . The ultimate data must be given to us by a non-reasoning process—a self-knowledge of that which is in our consciousness."

So far I have been dealing mainly with the implications of Eddington's philosophy, taking him as an example of one of the most thoughtful of the explorers into the innermost recesses of mathematical physics. But fortunately Eddington himself has developed his own thesis in these directions, in the Swarthmore Lecture which he gave last year to the Society of Friends, and which is published as a small book called *Science and the Unseen World.* Hence, if I have unintentionally misrepresented him in any respect, the misrepresentation can be corrected by reading that comparatively easy and remarkable pronouncement.

Instead of Eddington as an example of an idealising mathematical physicist, I might equally well have taken Whitehead, from whose notable book, *Science and the Modern World*, I have elsewhere made considerable extracts and appreciative comments (*Proceedings S.P.R.* for June 1929). But Whitehead does not write with the same clearness and vividness as Eddington does; and though I find myself even more in agreement with Whitehead than with Eddington (or imagine that I do, for I don't always understand him), he will hardly, I think, make the same impression. Whitehead and Eddington do not always take the same point of view, they sometimes usefully disagree. Eddington agrees much more with Einstein than Whitehead does. Yet their disagreements do not seem to me fundamental; and Whitehead carries the matter even further, in some respects, in the idealistic direction. He does not limit physics to pointer-readings. He is willing to extend the methods of science beyond their ordinary scope. He looks upon the universe as a kind of *organism*, and detects attributes akin to those of life or living organisms everywhere. And yet his views are mainly based upon mathematical physics, which is his own subject, though he seems well acquainted with the history of philosophy too, and he will perhaps be more acceptable to professed philosophical readers than Eddington is. He deals, in fact, less with the intricacies of physics, and employs more consistently the language and methods of philosophy. In comparing Eddington with Whitehead, the interest for me, and I hope for others, is that, though these two eminent and learned men differ in detail, their conclusions are ultimately rather similar.

Some Biological Contributions towards Mental Activity and Guidance

And here, quite recently, is a contribution from the side of biology, or at least from the side of experimental psychology. On May 14, 1929, the Herbert Spencer Lecture was delivered at Oxford by Dr. Charles S. Myers of Cambridge, under the title "Psychological Conceptions in Other Sciences". In this he accepts most of the conclusions of modern physics, realises the probable truth of both guidance and mechanism, and sustains it by arguments derived from experimental psychology He claims that psychology emphasised relativity long before physics attended to it (though that is something of an exaggeration). Physical theory, he says, is fast abandoning its former notions of substance and absoluteness. Mechanical determinism is limited.

> "The once striking characteristics distinguishing Matter from Mind are fading rapidly. Mind appears to be no more 'unsubstantial' than Matter; Matter to be no more 'predictable' than Mind. To account for the Evolution, the history and conduct, of the Universe, or of any organised individual within the Universe, whether relating to Mind, Life, or Matter, not only mechanical principles but also a certain adapting, selecting, guiding activity must ultimately be included among the *First Principles* of Science."

Dr. Myers' pamphlet is a kind of supplement to Eddington and Whitehead, though on a much smaller scale. It is short, and I need not make extracts from it. He points out that there is a working machine in evolution, but that the machine needs directive guidance, either predetermined or continually selected and adapted. In addition to conservation of energy and conservation of matter, he emphasises what he calls the psychologist's peculiar principle, "conservation of self", or, more generally, conservation of the individual system; which makes for preservation of pattern, endurance of identity, and unity and continuity of the individual, throughout its history. The biological struggle for existence is essentially a psychical struggle, and is an illustration of the conservation of self. This principle he finds stereotyped in the patterns of animal and vegetable life, and wishes to invoke it also to account for the distinctive patterns and types of movement and configuration observable throughout the physical, chemical, and stellar worlds.

He points out that the constituent elements into which science tends to analyse reality come much later, and are really more complicated, than the simple wholes which are presented to experience and consciousness. It is not grammatical rules which make a language; the language comes first, and the rules are analysed out afterwards. A sonata of Beethoven is not compounded and built up as a combination of elementary scales and chords; the music came as a whole into the mind of the composer. It can be analysed afterwards, and attention can be directed to the utilisation of fragmentary scales and chords by the expository skill of Sir Walford Davies.

Dr. Myers does not use this illustration, but I think he would agree with it. The universe is given to us as a homogeneous whole, and its heterogeneous elements are obtained by differentiation. It is not that the whole is aggregated or built up out of its elements: the elements are as it were carved out of the pre-existing and evolving whole; they are not combined to produce it.

> "The old view expressed by Spencer, and not yet wholly extinct among biologists, was that consciousness has arisen as an 'epiphenomenal' product of living matter when physiological processes became too complex to work automatically. The modern psychological view is the direct converse of this—namely, that consciousness, however primitive, fulfilling, however feebly, the function of orderly direction and purpose, is primary, and that it has grown by distillation, differentiation, and restriction, to narrower, more dominating, higher levels within the organism."

He then makes a quotation from an admirable but neglected physicist, John James Waterston,

> "who in 1843 gave expression to the view that the phenomena of Mind will one day be employed to throw light on the phenomena of Matter . . . [that organisation is the striking feature and] that if molecular philosophy is ever destined to advance into the region of [biological] organisation the phenomena of perceptive consciousness will admit of being applied to illustrate the physical aspect of the elementary processes of matter".

Organic phenomena cannot be explained by mechanical arrangement without some organising power which Waterston called "molecular adaptations".

Something of the same sort is found in Eddington, where he claims that the design and organisation, rather than the resulting elaborated products, are the important things; he directs attention not to the energy alone, but to the organisation of the energy.

> "There is no doubt that the scheme of physics, as it has stood for the last three-quarters of a century, postulates a date at which either the entities of the universe were created in a state of high organisation, or pre-existing entities were endowed with that organisation which they have been squandering ever since. Moreover, this organisation is admittedly the antithesis of chance. It is something which could not occur fortuitously."

The external world is not a mere assemblage of myriads of particles, not a mere collection of atoms, ether waves, and the like, going about their business.

> "There is a side of our personality which impels us to dwell on beauty and other æsthetic significances in Nature, and in the work of man, so that our environment means to us much that is not warranted by anything found in the scientific inventory of its structure.

> An overwhelming feeling tells us that this is right and indispensable to the purpose of our existence."

I might make somewhat similar citations from General Smuts' important book on Holism, but there is little need to call attention to this, or to the notable series of Gifford Lectures by the physiologist, Dr. John Scott Haldane, published under the title *The Sciences and Philosophy*, a book that contains much with which I sympathise. So does Professor McDougall's *Body and Mind*; and, indeed, many other recent books might be referred to, the writers often differing in detail, but on the whole agreeing in their main contentions, as contrasted with those of the materialistic philosophers of last century. I mention these books in order to illustrate and justify my contention that different thinkers from different points of view are converging on some kind of guidance, some kind of mental activity, some kind of organising power, some rational and predetermining influence, not only in the works of man, where it is conspicuous, but in the works of nature too. I fully expect that this guiding power, whether we call it life or mind or entelechy or what not, this directive and organising tendency, has physical mechanism associated with it, and that it is through its concealed physical mechanism that it is able to operate on the sensible and tangible world of atoms and electrons. On this subject I have ideas, and they are not vague; but they have not yet risen to the dignity of a physical theory, and I doubt if they should be intruded on your notice. I have been asked to go "beyond physics", but I have no wish to separate myself from a physical base, for I conjecture that physics and psychics are continually and permanently interlocked. And yet the physical mechanism, say of thought or mind or directing control, is so concealed and at present so unknown that one cannot make definite statements about it, even of the highly abstract and yet quantitatively definite kind which would ensure attention from the workers in modern physics.

A Transition Period

Perhaps it will be best for me to remember that I have been asked to give this address as a nineteenth-century physicist, one who was brought up in the atmosphere of Stokes and Kelvin and Tait and Maxwell and Rayleigh, one who co-operated with FitzGerald and Poynting, and regards with admiration the work of Sir Joseph Larmor and Sir J. J. Thomson, one whose foundations were laid before any of the modern developments had shown above the horizon, one who had first regarded the theory of relativity and the quantum theory with a kind of apprehension and dislike, and yet who has been led by the compulsion of fact to realise that in these ways a broader, deeper, and more intimate perception of the processes of nature was obtainable. I perceive or admit that the old mechanical or engineering ether is extinct, but hope that a new far less familiar but much more accommodating ether is rising above the horizon, and that radiation or ether waves of one kind or another may ultimately turn out to be the most enduring reality in physical existence.

What the great men that I have cited would have thought of all these new developments I cannot say. To some extent, and in their own time, they were rebellious; and yet unconsciously they were pioneers even of the new knowledge. Larmor and J. J. Thomson are still with us, and can speak for themselves. That they acquiesce entirely is not to be expected; that they are watchfully critical is certain. The abandonment of old methods may seem to some of the younger workers more complete than it is. The method of working in nothing but mathematical abstractions, with no physical image or concrete ideas to catch hold of, may turn out in the long run impracticable. That progress can be made at all in such ways is astonishing. There may have to be some return to the old paths. Professor Whittaker of Edinburgh is working in that direction; and so,

I think, are the others that I have mentioned—with what success remains to be seen.

It is well to remember, for it is apt to be forgotten, that many of the theoretical results now attributed to Relativity are based essentially upon the old nineteenth-century physics. That the mass of an electron is not constant, but is a function of its speed, was worked out long ago both by Heaviside and by J. J. Thomson. And as soon as the electrical theory of matter was established, the same conclusion applied to all matter. The variation of mass is often called for brevity a "relativity correction", and there is some excuse for that term; but it is not quite fair historically, for the variation, precisely as observed, is a consequence of orthodox electromagnetic theory.

The equations of Clerk Maxwell, moreover, remain in possession of the field, and the deductions that have been made from them continue valid. The Lorentz transformation, on which so much depends when we shift from one system of moving axes to another, was devised in final form by Larmor, being supplemented and made reciprocal by the additional factor of the Fitz-Gerald contraction. All these things have gradually evolved, and it is difficult to specify any particular origin. Brilliant use has been made of them by the relativists, who have exercised their selective power in emphasising the absoluteness of certain things as compared with others. The search for absolutes is a very legitimate quest; though we may have to admit that the quest originated in certain hypothetical postulates, which though they are now justified and partially verified by the results obtained from them, have never been finally and completely verified experimentally. One such postulate is the unspecifiability and meaninglessness of locomotion through the ether; another is the absolute constancy of the velocity of light as measured by all observers, no matter how fast they themselves may be moving towards or away from the beam. And though this latter surprising postulate has become attractive, and if true suggests something very important about the constitution of the ether of space, yet it does not seem wise to accept either that or the other too gratuitously, or to shut the door to the need for actual verification. It is never wise to shut the door on any possible experiment; at present the difficulty is that a crucial experiment hardly seems possible.[1]

[1] Motion of the source could have no effect on speed, on any reasonable view, so long as light consists of waves alone; but common sense would say that motion of the observer to meet the light must hasten its arrival. This, however, would contradict one principle of relativity, which insists that so long as source and observer move together nothing can be observed. I wonder if the skill of Prof. A. A. Michelson can devise a test experiment at Pasadena. The difficulty is to dispense with a return journey. There could

Speaking as a nineteenth-century and conservative physicist, and knowing that my remarks will be received with some contumely by the Younger School, I dislike the emphasis which is laid upon observations and measurements made by observers travelling at high speed; a mode of argument presumably adopted for popular consumption, but with an element of absurdity in it. To say that an observer flying at 1,000 miles a second, or for that matter even walking at four miles an hour through a laboratory, can make measurements just as good as one who sits down to his observation and makes any necessary corrections afterwards, is hardly in accordance with common sense. Measurement is not simply reading a scale or dial, it is a mental deduction, supplemented by calculation, using the scale-reading as an indicator. The relativist's "observer" doubtless takes his instrument with him, the instrument is part of the observer, but he is supposed to observe and measure details of things moving at enormous speeds relatively to himself. The attempt of an expositor to make things intelligible is conducted under difficulties; probably he is not misled by them himself, but the public is apt to be confused.

The most recent methods of abstract symbolism,

be no Doppler effect when source and receiver are both clamped to the earth, but change of speed there might be, down an ether stream. Locomotion through ether would then have a meaning.

in terms of tensors and matrices, are of no use to a mind that requires something more pictorial and conceivable. The dweller on the ground can admire the evolutions of a skilled aviator, without having any hope of taking part in them.

To illustrate the extremely abstract treatment of nature by the modern mathematical physicist, even in its simpler forms, I will refer to an article in *Mind*, expounding a modern view about the nature of matter. In the issue of *Mind* for April 1920 Eddington gave a paper on "The Meaning of Matter". The highly abstract mode of regarding matter, there advocated, expresses it in terms of the tensor $G_{\mu\nu}$ and the curvature of space. Eddington's expression for it, at that date, was $G_{\mu\nu} - \frac{1}{2}g_{\mu\nu}G$; in other words, he claims that matter is a variety of space-curvature and nothing else. It isn't that matter is something which curves or warps space, but that matter *is* the curvature. Furthermore, he pointed out in his book *Space, Time and Gravitation*, page 192, that some approach to these views was suggested with marvellous foresight by that remarkable genius W. K. Clifford, forty years ago.

> "Whilst other English physicists were distracted by vortex-atoms and other will-o'-the-wisps, Clifford was convinced that matter and the motion of matter were aspects of space-curvature *and nothing more.* And he was no less convinced that these geometrical notions were

> only partial aspects of the relations of what he calls 'elements of feeling'. 'The reality corresponding to our perception of the motion of matter is an element of the complex thing we call feeling. What we might perceive as a plexus of nerve-disturbances is really in itself a feeling; and the succession of feelings which constitutes a man's consciousness is the reality which produces in our minds the perception of the motions of his brains. . . . The theory of space-curvature hints at a possibility of describing matter and motion in terms of extension only . . . the future theorist has to build up the world as best he may.'"

But I must not waste time in generalities. Let us see whether there are a few points on which I can indicate, in the more immediately physical parts of Eddington's brilliant 1927 exposition, where I should like to emphasise his contentions, and where I should like to put in a cautionary word.

Light as Messenger

A minor point on which I feel inclined to raise a small question is the singular importance which philosophic relativists attach to the speed of the quickest messenger available, in other words to the velocity of light, as something fundamentally affecting their notions of space and time. For instance, take Eddington's diagrams in his Gifford Lectures, pages 41–8, of what he calls "me", travelling along the stream of time, from the absolute past towards the absolute future, between parts of space which are "elsewhere". (It is really intended as a four-dimensional diagram, in which only one dimension of space is represented, the other one depicted being the time dimension.) The point where "me" is, is called "here-now", and a horizontal line drawn through that point represents "now", or simultaneity in other parts of space. But inasmuch as we have no instantaneous messenger to tell us what is happening in other parts of space, he very soon discards this "now" as meaningless, thus complicating the idea of simultaneity, and uses instead a slant line labelled "seen now", the slant corresponding to the velocity of light. There is a pair of such lines, one with an up slant, the other with a down, and the result is to partition off a region of space which might be called the "inaccessible" region; this he calls "absolute elsewhere", meaning that by no device can we ascertain what is going on there, for we have no means of getting at it unless we or some other messenger can travel quicker than light.

I am ready to admit that the transmission of anything substantial or energetic quicker than light is impossible; but I feel doubtful whether "seen now" ought to have so important a philosophical significance. For if our vehicle of apprehension were, say, Sound, we should draw a slant line just a million times steeper, and label it "heard now". This line would obviously be of no philosophic importance; and yet if nothing could travel faster than the velocity of sound, it would at first sight seem to have as much philosophical significance as the other. The difference, however, is that sound is an affection of matter, while light is an affection of space or ether, and therefore presumably is much more fundamental. It may be that relativists are right in instinctively attaching so much importance to the velocity of light, because the velocity which goes by that name seems to be a fundamental velocity in the universe, characteristic of the ether, in or of which everything consists. Its *primary* significance is probably the constitutional turbulent or rotational velocity of the ether, and only incidentally or in secondary fashion does it represent the speed with which that fundamental substance conveys waves.

It is this possibility, which I agree is very important if true, that leads me to criticise the diagrammatic procedure only in tentative and apologetic fashion. But otherwise the mere seeing of a phenomenon that has occurred at a distance, possibly a century after it has occurred, though interesting enough in itself, hardly seems to me a valid reason for questioning the meaning of simultaneity, and using the velocity of light to partition off three regions—one the absolute past, which is over and done with and unalterable; another the absolute future, into which alone we are able to go when the proper time has arrived; and a third region, which is wholly inaccessible to any of our senses, about which we can gain no information, a region of space which is absolutely and hopelessly "elsewhere" For I would venture to say that though occurrences in that region are physically inaccessible to us, and can only be known when they have crossed the "seen now" line and become past, yet psychically they are accessible. We can *think* of phenomena before they have occurred; and, indeed, we might make preparations for observing them when, later on, they are going to cross the diagrammatic line of visibility and seem to us to occur. In other words, I suggest that *thought* can penetrate even the inaccessible region, though it must be admitted without any physical justification or observable result until the proper time arrives. We shall find that waves do exist that can travel quicker than light, though whether we ever unconsciously utilise them is a question as yet unanswered.

My objection to philosophic emphasis on physical "seeing" may be taken as one of a more general kind, namely, that Relativists seem inclined to over-emphasise the laboratory method of observation in a curiously unpractical and imaginative fashion. The appeal is constantly being made to an "observer"; but the observer is apt to be placed under ludicrous conditions. And though his procedure seems able to be reasoned about, and deductions drawn as to what is absolute and what is not, yet the experiments are so imaginative that I doubt if they will stand the stress which is being put upon them. If we granted the feasibility of high-speed observers, they would doubtless be affected by all manner of FitzGerald contractions and sluggishness of clocks; but I doubt if such hypothetical observers are seriously admissible.

The relativist contention is that without knowing it we may be such observers ourselves. Motion through the ether is said to be meaningless. It is easy to *imagine* enormous velocities, and who knows but that we might be affected with some such velocity? That is where I join issue. I do not think velocity through the ether is meaningless, though I am bound to admit that we have no means of measuring it at present. I regard it, however, as a legitimate effort of science to try to ascertain that velocity. I do not think it legitimate to say that the Ptolemaic system, in which the stars revolve round the earth once a day and once a year, is just as right as the Copernican system, where these motions are attributed to the earth. We may not be able as yet to measure motion except with reference to other pieces of matter, but some day I hope that it may be measured with reference to the ether; or if not to the ether itself, then perhaps to the rare cloud of interstellar matter which recent advances have detected, distributed throughout space, with an average motion which relative to the ether may be assumed nil, so that it can serve as a material standard of reference.

Apart, however, from any material standard, I should take the ether, or in other words space itself, as stationary as regards locomotion; and every velocity relative to that I should call absolute velocity. The relativist mode of reasoning is doubtless legitimate, for it is legalised and justified by its results; it does enable deductions to be made, but it seems to me a mathematical device rather than something of philosophical importance; the kind of thing that Eddington has elsewhere called "a dodge, and a good dodge too". I should hope, however, that the progress of physics will render many things, such for instance as the FitzGerald contraction, definite and ascertainable, and not leave them at the mercy of imaginary and handicapped explorers, who at present are supposed each to claim that his own observation is as good as any others. The Relativists have already detected certain absolutes on which they lay stress, quite rightly. I would venture to urge them to keep an open mind for some other absolutes, which at present they seem to me too apt to deny; absolute locomotion through the ether being one, and perhaps the chief of them.

I have no fault to find with the absolutes already discovered. The constitutional velocity c is a great discovery, and its importance is emphasised by the fact that it enters into the composition of velocities of quite ordinary things, so that we cannot calculate a resultant velocity by simple addition without taking c into account. And the way in which this method of composition of velocities accounts for the Fresnel-Fizeau speed of light in moving water is very striking and exhilarating, and seems to justify the Larmor-Lorentz transformation to a surprising degree. In fact the temptation often is to throw in one's lot whole-heartedly with the Relativists, whose mode of dealing with things is so successful, and to withdraw any cautious and conservative opposition. But I think it would be a pity to do that prematurely, since it tends to shut the door on possible test experiments which can some day be actually made. Many attempts have been made to ascertain the speed of matter through ether, or absolute motion in space; and though they have all hitherto failed, I don't want it to be assumed too readily that in the nature of things they are always bound to fail; which is one of the chief relativity postulates.

To the quantum I have no sort of objection: only I think it may be rationally accounted for, some day, by regular mechanics. It is a remarkable and most interesting discontinuity due to the discontinuity of matter, and possibly to some unknown rotational boundaries existing even in the ether. It dominates the interchange of energy between ether and matter, and lies at the root of all radiation processes. The discovery of the discontinuity of electricity was most important; the discovery of the quantum is of perhaps equal importance. At present, in dealing with the quantum, we are in the midst of physics and yet seem to be opening up an avenue into a region beyond physics, at any rate beyond the material aspect of physics. Similarly the discoverer of the electron, himself, has not hesitated to affix to a lecture which he gave at Girton last year the challenging title of "Beyond the Electron". Indeed, the nature of the electron, which at one time seemed so simple, though it was never pretended to be "known", is now one of the chief battlegrounds of modern physics, and its partial resolution into waves is one of the most hopeful signs of the resolution of matter into some form of ether vibration or circulation, some form of constitutional periodicity in space.

Entropy

I shall hope to return to that, but meanwhile let me call attention to the remarkable contention of Eddington about the Second Law of Thermodynamics and the doctrine of Entropy. He considers this "second law" the most fundamental law of nature, to which no exception can even be imagined; and he uses it to emphasise our conception of the progress of time, which must always proceed in the direction of increasing entropy, or energy dissipation. Time, he virtually says, differs from space in having direction, an arrow-head, like a river rather than a road. Mathematically time can be thought of as going either forward or backward, a mere difference between plus and minus; but actual time can only go forward. In this it differs from space: "right" can become "left", but "after" can never become "before". Any attempt to run time backwards would only result in hopeless confusion.

This used not to be the doctrine of W. K. Clifford. I remember his saying that if at any given instant every atom in the universe were reversed in direction, things would begin to unhappen, and the course of evolution would retrace its steps. Only he admitted that this procedure would be unstable, whereas the straightforward course was stable, and not liable to interference by any small disturbance, as the unstable course of events would be.

Eddington, however, would disagree with that. His test of time's progress is the increase of entropy. He regards a universal decrease of entropy as impossible or even inconceivable, even for a moment.

Now entropy is not a simple idea with which we are familiar, like heat and temperature for instance. It originated in thermodynamics, and has become useful in engineering, but it is not one of our ordinary conceptions, and therefore feels a little difficult. It is defined in thermodynamics so that a change of entropy means the change of heat in a body divided by the absolute temperature of that body. When heat flows uncontrolled from a hot body to a cold, the entropy of the system is greater than before: the cold body gains more entropy than the hot body loses, by the transfer of a given amount of heat without any work being done in the process. This kind of flow is frequent and automatic, it needs no regulation or control. When a thermal operation is under human control, as it is for instance in the working of a perfect heat-engine, there need be no increase of entropy. Heat is taken from a hot body at one temperature, part of it is converted into the energy of mechanical movement, say of pistons and flywheel, and only the balance is given to a body at low temperature. The ratio of the heat taken to the temperature of supply may be equal to the ratio of the heat given up to the temperature of withdrawal. In that case the total entropy remains constant.

But it is equally possible, and indeed much commoner, for no engine to be used, for no organised motion to be produced, and for the heat from a hot body to be conducted down to a cold body, so that the ratio of heat to temperature, in other words the entropy, is increased. This process, which Kelvin called the dissipation of energy, is continually going on. Hence it is said that the universe is running down, and that we have no means of reversing the operation and winding it up again. This is Eddington's test of the progress of time, viz. to watch an irreversible operation.

By suitable devices entropy can be kept constant, and in that case the operation is reversible. But once an irreversible operation has been performed, there is no remedy, it cannot be undone. Time has gone on, and it cannot go backwards. Entropy has been irremediably increased.

In order that an operation may be reversible, the energy concerned must be under control. It must not be allowed to run riot. Any random operation is irreversible. A fire or a flood under control is useful, it could be employed to produce organised motion, say by a steam-engine or a water-wheel. But out of control it can only do damage. The energy is not lost, it is only unavailable, when it ceases to be under control. Milk or wine spilt on a sandy beach is all there, but it is irrecoverable. The chemicals used in warfare are hopelessly wasted. An organised army of workmen can erect a building or a bridge; a disorganised mob is a source of danger. Everything depends on organisation and control.

This has long been known, but Eddington illustrates it very luminously by what he calls the operation of "shuffling". Given an orderly pack of cards, it may be hopelessly disorganised by shuffling, and no amount of shuffling will bring it back into order. Many of the processes in nature thus result in greater disorganisation; and according to Eddington the irreversible disorganisation measures the entropy. Entropy is disorganisation. It is easy to break an orderly arrangement down, but not so easy to build it up.

Yet it can be built up. Not by random and unintelligent processes truly: a mob of monkeys playing on a million typewriters will not compose a volume of poems. The only way to restore order is to apply the activity of mind. The demolition of a building seems irreversible, but by taking thought it might be restored. A shuffled pack *can* be rearranged; indeed, this is done in the game called Patience, when successful. Shuffling, as Eddington luminously says, is "an absent-minded operation". One could imagine the shuffling of a conjurer which should leave the cards in any order that he wants: that would not be an absent-minded operation. Mind is essential to organisation, and organisation or reorganisation is a natural result of mental activity consciously directed to a foreseen end.

Influence of Mind

Hence I very much doubt whether the second law of thermodynamics has the dominating position which Eddington claims for it—at least not when we go beyond physics. He says that entropy can never be decreased; he means, of course, by the inorganic operations of physics. He would surely admit that waste can be checked by mental operations, that is by the operations of life and mind. They are able to take things under control, and thereby restore order out of disorganisation. But this is surely very important. Life and mind bring in an element which has not yet been incorporated into physics. The era is approaching when that element ought to be incorporated. The psychic element has been too much ignored; the effort is made to limit physics or to define it so as to apply only to metrical observation and to unorganised and random processes. Live things are excluded from the laboratory except when they are mere observers; they introduce perturbations which cannot be predicted; they have an element of spontaneity and free will. There is no telling what a spider is going to do next, or for that matter a charwoman either.

Such agents are requested not to interfere with the instruments, nor even to dust one's papers.

So long as we take the universe as unaffected by life and mind, the increase of order, the production of cosmos out of chaos, may be impossible. Random or absent-minded operations are useless, they only increase the shuffling; and once a thing is shuffled into complete disorder a final mess is made of it, from the point of view of unaided physics. Order cannot be restored by absent-minded action, however long continued, except momentarily by remote chance; and no bell will be rung to indicate the happening of that chance.

But my point is (and after all it is a platitude) that life and mind are not excluded from the universe, and that therefore it need not always be running down into disorder. It may all the time be more under control than we know. At any rate the operations of life can take the random materials of carbonic acid and water and build them up into an apple-tree. Life confers upon the assemblage a specific and even beautiful form, with the marvellous possibility of continuing that organisation for any length of time. This does not mean that time has gone backwards, or that time's arrow has been reversed. It means the introduction of a biological and teleological element into an otherwise complete scheme of physics.

I suppose Eddington would say that he never meant physics to include biology and psychology. He regards physical science as a limited aspect of the universe; he claims that its operations may run round in cycles, free from consciousness, under no vital or mental control, and that under these conditions entropy or disorganisation would always increase. That is plausible, though even that (for the moment) I venture to doubt. At any rate I would ask him whether the formation of a crystal would not be troublesome to an observer who thought to trace the direction of time by the increase of disorganisation. The random molecules in the solution achieve an organised solid form, and they do this by their own physical properties. I do not press the point, for they are not under the control of life and mind, but it would seem that even in the inorganic world a statement of the doctrine of entropy has to be carefully guarded; and I do claim that directly we introduce mental or even vital activity the whole law is upset.

I can imagine Eddington retorting that the physical activities of life and mind are only rendered possible because of the blaze of running-down energy emanating from the sun. I never meant, he might say, that you could not wind up a clock! Solar radiation winds up the water from the sea; there are all manner of winding-up processes even in inorganic nature, but they are all local and partial recoveries, only rendered possible by more running down somewhere else. A water-ram may pump water to any height provided a much larger weight of water descends down a smaller height. All this is ancient and orthodox doctrine. I admit, of course, that we are constantly making use of solar energy, and so is vegetation. I never suggested that life and mind could generate energy, I only claim that they guide it, and therefore interfere with the process of running down. Their function is mainly that of winding things up, organising things, sorting things, introducing order. We reclaim soils, organise farming operations, building operations, construct time-tables, and carry out prearranged plans; we regulate things of all sorts. In so far as life acts at all, it is an organising and directing or guiding power. Well, I want to recognise that on a cosmic scale.

Maxwell called attention to this function of intelligence long ago. The passage of heat from hot to cold is generally considered irreversible; the passage of atoms from a crowded half of a reservoir into a vacuous other half, when a partition between them is withdrawn, is also irreversible. Never again will the particles be sorted all into one half. The improbability is practically infinite. There is nothing to prevent all the inhabitants of London from congregating into the West End, but they are not likely to do it. Yet they might if there were some sufficient inducement. There is no obvious reason why the inhabitants of London should prefer underground passages to living in free air on the surface, but during an air-raid many of them did. And Maxwell pointed out that the diffusion of matter—and therewith the dissipation of energy—might be reversed by the intelligent activity of an imaginary being whom he called a "demon", who could witness the actual molecules in their flight, and deal with them individually. Such a being might control the molecules, as say Tilden or Suzanne controls a tennis-ball. Life is a guiding and directing principle which can undo the random operations of inorganic material, and produce results of unexpected and perhaps inexplicable beauty. Mind arranges the myriad tones of an orchestra, and lo, a symphony! Something has converted part of the turmoil of a gaseous nebula into a landscape. There is no increase of disorganisation about a symphony; I am not sure that there is about a sunset. No one can rationally claim that a rose is the product of random forces. Some people may try to think that a planet is such a product, but on the whole they must realise that they fail.

It is needless to multiply illustrations. The meaning of what I am saying is clear enough. The time has come when we ought to try to bring life and mind into the scheme of physics, and we shall not fully understand the nature of the physical world until we do. But now comes the perennial difficulty: what must be the nature of these entities if they are to interfere with and operate on matter? How can things of one category act on things of another? The first step is to reply that conspicuously they do, whether we understand the operation or not. And the second step is to make some attempt to understand how they do it.

Vitality and Mechanism

We find that life is always associated with a material organism so far as it is known to us at all. We find that there is mechanism underlying apparently every mental operation. A physiologist seeks to trace the mechanism throughout. Well, let us accept that idea, and see if we can find a physical mechanism available for life and mind in general, whether it be associated with matter or not. For many things there are which act upon matter and yet are not material. Electricity, magnetism, and light are affections of space or ether, and yet they act on matter.

I have several times of late tried to suggest the idea that life and mind may be perhaps, not exactly affections of the ether, but may have some etheric or spatial concomitant inevitably associated with them, which enables them to exert an influence on the material world. My hypothesis is that physics and psychics are interlocked. Not that mind is necessarily associated with matter, but that nevertheless it is associated with something physical—a much broader term than material—something which may properly be dealt with in physics when the scope of that subject is sufficiently enlarged, and when the discoveries now in process of being made reach their fruition. In such speculations I am going beyond physics, but it is just beyond physics that I was asked to go.

I am now going back to some of the latest developments in modern physics to see if there is any physical agent which can be used for vital or mental purposes. The search is for something in the physical world which is not energy but which is capable of guiding or directing energy. I must conduct the search through an apparent digression.

CHAPTER V

THE PHYSICAL PROPERTIES OF GROUP WAVES

OBSERVANT people must have occasionally noticed that when they walked past two sets of palings at a moderate distance apart they saw, in addition to the row of palings nearest them, as it were a ghostly set of palings much broader than the real, which moved more or less with them. These subjective appearances are easily explained; they are due to the alternate coincidences, alternate agreements and disagreements, of the position of the two sets of palings, as seen by the observer. Sometimes the palings coincide in the line of vision, and then the interspaces are clear, we see through both sets; in other places the pale from the one set blocks up the aperture of the other, and so vision is obstructed, and you get a dark patch. It is these dark patches, occurring at regular intervals, that constitute the appearance of a ghostly travelling set of palings. Sometimes when two fences are on either side of a raised path on an embankment near which we are motoring at a lower level, the appearance is very striking and peculiar. The elevation of the objects then enables the sky to be seen through them when they coincide, and the periodic obstructions are conspicuous.

These phantasmal appearances have a speed of their own, dependent partly on the frequency of the real palings, that is on the number of them per yard, and partly on the motion of the observer. The interval between the phantasmal appearances, or what may improperly be called their wave-length, depends on many relative considerations.

The same sort of thing can occur under many different circumstances, but it always involves two sets of regular objects with either really different or apparently different periodicities, the differences being small. In the laboratory, when two separated portions of a beam of light are made to coalesce upon a white screen, the alternate agreements and disagreements in the wave motion appeal to the eye of an observer as what are called interference bands. These bands are much coarser than the waves of light, and can be made quite broad; their width, in conjunction with the distance of screen from separator, depends on the spacing out or "length" of the constituent waves, and indeed enables those waves to be measured. The waves of light are progressive, but the coincidental or interference bands are stationary. They represent a redistribution of energy in space: the result is a periodic simultaneity.

A similar phenomenon can occur in connection with sound; but our sense of sound is periodic in *time*, and accordingly we get successive alternations of sound and silence, well known as beats. All these phenomena may be called generically "beat" phenomena. Sometimes the coincidences are real, so that it is a real distribution of energy, as in the case of light and sound; sometimes it is merely an optical illusion depending on the observer, as in the case of the palings. (When both sets of palings happen to be identical in width, or what corresponds with wave-length or frequency, the relative or apparent difference between them is due to their different distances from the observer.)

A similar sort of thing can be observed in the case of water waves, or with any waves whose velocity is dependent on wave-length. In that case the coincidences determine some special feature, perhaps a bigger wave than usual, or some other mark of identification. And such coincidental waves are known as group waves; they have a velocity of their own, which is called the group velocity. This is slower than the true wave velocity, having indeed half its value in the case of water waves: the true waves may be seen passing through the group. I believe that on the open ocean patches or lumps of these group waves can be seen, preserving a sort of identity, and having a secondary or derived speed of their own. The theory of such group waves was worked out most clearly by the late Lord Rayleigh.

They are a recognised phenomenon with many applications.

Energy Transmission by Waves

Two things, two different processes or functions, are involved in all wave motion: the transmission of energy, and the propagation of wave-forms. This double aspect always exists, though in light and sound it is obscured because the speed of energy-transfer, in those special cases, happens to be equal to the speed of wave-propagation. In a *dispersive* medium, since wave-velocity therein depends on wave-length, wave-energy must travel at a different rate from wave-forms. It appears to be a general characteristic of wave-motion that energy is conveyed solely by the group waves. When energy is not conveyed, it does not mean that the medium possesses no energy, but that it is not able to transmit it. In other words, the speed of group-wave transmission may be nil; although wave-forms are travelling freely.

It is natural to be surprised at the existence of wave-forms which convey no energy and yet travel at a great pace. I am glad to recall attention to a clear-sighted paper communicated to Section A of the Plymouth Meeting of the British Association in 1877 (see *Nature*, vol. 16, p. 343) by that eminent engineer the late Professor Osborne Reynolds. He maintains, in that paper, written so long ago, that the two functions, propagation of waves and transmission of energy, are independent of each other; and he instances a simple case of no energy transmission, namely a row of disconnected pendulums, all alike. By starting the pendulums oscillating, one after the other regularly, it is easy to get a wave-form advancing with a speed exactly proportional to the wave-length. In such a case the frequency is definite—the time period is that of the pendulum oscillation—but the wave-length is optional, so the velocity is optional too. There is energy in the oscillation, but none is transmitted; the velocity of transmission of energy is zero. The velocity of propagation of the waves is arbitrary; it is as usual λ/T, and T depends entirely on the length of each pendulum, but λ is undetermined. The medium surrounding the set of pendulums takes no part in the action; it has virtually no properties, i.e. no relevant properties. One may make some of the pendulums swing more than or differently from the others, and thus get a group, but the group does not travel: it and its energy remain steadfast as regards locomotion.

If we introduce some elastic connexion between the pendulums, that will represent a medium uniting them, and now energy must begin to be transmitted. Energy belongs to a group of waves and now travels with the group. The theory was clearly exposed by Lord Rayleigh, and can be found either in his writings (Appendix to Vol. I of his *Theory of Sound*) or in Professor Horace Lamb's *Textbook of Hydrodynamics*, chap. vii. Two component waves being written as $\sin k(x - vt)$ and $\sin k'(x - v't)$, the speed of the combination, or group system of beat or interference waves, can be expressed in any of the equivalent forms:

$$u = \frac{kv - k'v'}{k - k'} \text{ or } \frac{dkv}{dk} \text{ or } \frac{dv}{dN} \text{ or } \frac{d\left(\frac{1}{T}\right)}{d\left(\frac{1}{\lambda}\right)} \text{ or } \left(v - \lambda\frac{dv}{d\lambda}\right)$$

whereas the speed v of the waves themselves is $v = \frac{\lambda}{T}$ or $\frac{\nu}{N}$ simply.[1]

[1] In justification of the above: many schoolboys have learnt by heart, or at least have been taught, that "the sum of two sines is twice the sine of half the sum into the cosine of half the difference"; so the resultant wave, obtained by adding together the two component waves of the last paragraph, is

$$2 \sin\left(\frac{k + k'}{2} x - \frac{kv + k'v'}{2}.t\right) \cos\left(\frac{k - k'}{2} x - \frac{kv - k'v'}{2}.t\right)$$

which, when the constants differ only slightly, simplifies down to

$$2 \cos \tfrac{1}{2} dk\left(x - \frac{dkv}{dk}.t\right) \sin k(x - vt)$$

where the small difference $k - k'$ is written as dk, and $kv - k'v'$ as dkv. This means that the superposition of two neighbouring component sine waves yields a sinuous wave

The importance of all this can be realised when it is held, as many are now inclined to hold, that all energy (not only potential but ultimately kinetic too) is really in the ether, and that it can only travel from one place to another by means of waves. Electrical energy is certainly transmitted in that way, and there is so great a unity about all forms of energy that the idea is growing that in no other way can energy travel except in the form of etheric waves. The difficulty, which at first seemed imminent, about slow locomotion of energy in some cases, as when matter regarded as a form of energy is brought to a standstill, is now overcome by the conception of group waves. Wave-forms must travel fast, and may travel extra fast, but then the energy, represented by group-waves, will travel extra slow. The remarkable thing is that the product of the speeds in ether is

the same in pitch and speed as either of the originals, but with a fluctuating amplitude. Alternate maxima and minima, ranging from 2 to 0, recur with regular periodicity, and these well-marked groups travel along with the velocity $\frac{dkv}{dk}$; as is shown by the slowly pulsating cosine part of the last expression. In the text this speed is called u.

To digress and anticipate for a moment: By distributing a number of values of k over a small range of frequency, called a "wave band", a complicated fluctuation of amplitude is represented, which can correspond with all that is necessary for speech or music in wireless transmission; and this fluctuation is precisely imposed on the carrier-wave by a microphone. The function of a receiver is to turn these groups or features of a complex ether-wave back into the sound they represent.

the square of the velocity of light. In other words, the speed at which free unmodified ether transmits both waves and energy is a mean proportional between *u* and *v*, the two wave-speeds under more sophisticated conditions.

Illustration from Broadcasting

Instances of group waves becoming perceptible to our senses, even though for some reason the constituent or component waves are imperceptible, are well known. One gets this sort of phenomenon in wireless telegraphy. The waves emitted by a station are much too slow to affect an eye, and much too quick to affect a telephone; but by affecting their amplitude with a superposed acoustic frequency, so that their intensity fluctuates sinuously, or—what is mathematically the same thing, see a recent footnote—by superposing two waves of nearly equal frequency upon each other, they can be heard in a telephone after rectification. For the beats or coincidences are much more widely distributed, and arrive as group waves with a combination frequency appropriate to the comparatively slow vibrations of sound. They may still be what we *call* rapid, and give a shrill musical note when received by proper apparatus; but the combined or heterodyned vibrations are only reckoned in thousands a second instead of in millions; and it is possible to reduce their frequency further till it agrees with those used in speech and music.

Beginnings of Wave Mechanics

The point is that waves too rapid to affect our senses can be made to affect them, at least when rectified through a telephone, either by varying their amplitude or by the coincidences of two or more sets; and another point is that the velocity of a group wave can be occasionally more or less under control. These facts, thus baldly and imperfectly summarised, have been used by Schrödinger for a novel theory of matter, what is called the Wave Theory.

If we attempt briefly to trace its history, it may be said to have originated in a study of the properties of the electron and the quantum theory of light. Nineteenth-century experiments on light substantiated a wave theory for that form of ether disturbance. An electric charge, or let us say an electron, seemed to be a singularity of a totally different character, one which had an individual permanence, with a controllable speed which could be watched; and it was discovered by J. J. Thomson to possess other properties which fitted it to become the recognised unit of the atom of matter.

And then (strange to say) twentieth-century experiments exhibited an astonishing connexion between these diverse things—the wave and the point electron. Waves of light of suitable frequency could eject an electron with an energy corresponding to the frequency, and either toss it up or toss it out of the atom. Similarly when an electron fell back into an atom it emitted a train of waves of frequency dependent on the energy which the electron had to spare, and which it must get rid of in order to settle down into its new quarters. The phenomenon is called photo-electricity; it lies at the basis of Bohr's theory of the atom, and is intimately connected with the quantum. For electron jumps are from level to level, or from step to step, with no fractions of a step permitted: there is nothing continuous about either the absorption or the emission of radiation. The mechanism, whatever it be, is more like a staircase than a slope; and the interchange of energy between ether and matter is always in quanta.

Then came De Broglie, who suggested a unification of the two interrelated things, the particle and the wave, so that the corpuscular and the wave theory of light might in a sense both be true. For a quantum of light is like a corpuscle, it has to be dealt with all or none, and yet light is composed of waves. Was the converse true, and was it possible that the electron was itself a wave of some kind—a group wave? Or are group waves associated with a travelling electron, their frequency or rate of vibration being dependent on its energy of travel? This latter question was put to the test, and has been answered in the affirmative by several experimenters, notably by young George Paget Thomson, the Professor of Physics in Aberdeen University, who has fired electrons of known energy against a photographic plate, and has shown that the point of impact is surrounded by diffraction or interference rings representing radiation of a corresponding frequency. The theory connecting frequency and energy, that is connecting the length of the waves with the momentum of the particle, as suggested by De Broglie, was verified. The waves when measured were of the right size, the size given by

$$\lambda = h/mu$$

His father, the present Master of Trinity, has carried the theory further, and has expounded the subject in his lecture "Beyond the Electron", already referred to; describing it in words in the body of the pamphlet, and giving a mathematical appendix at the end, all extremely worthy of study, and clearly set forth.

One remarkable outcome is that the difference between a wave and a particle is disappearing. We have to do with something that is neither, and yet shares the properties of both; something for which the name "wavicle" has been suggested. The constitution of an electron, notwithstanding its utterly minute size, is by no means so simple as we used to think it might be. In the outer orbits of an atom, a long way from the nucleus, an electron behaves something like a little sphere, revolving round a centre of force under astronomical laws. But an electron nearer the central nucleus, and therefore travelling at high speed and subject to stiff control, does not seem to be located at a point at all, but spreads out over the orbit like a succession of stationary waves. And the bare fact that for stability a whole number of pulsations must cover a closed curve—somewhat as they cover a violin string or an organ pipe—is sufficient to determine a succession of stable orbits, such orbits as had already been discovered in an atom by Bohr, with spectroscopic verification. The complete theory is rather complicated. All I want to insist on is that the difference between a wave and a particle is one of degree rather than of kind. A group wave of certain frequency may appeal to us as if it were a moving particle of given energy; and inasmuch as matter is composed of such particles, we may begin to generalise this into the suggestion of a group-wave theory of matter.

Schrödinger's theory carries the investigation further. Matter must obey the laws of dynamics: are group waves competent to obey such laws? The laws of dynamics are summed up in their most general form by the analytical methods of Lagrange and Hamilton. The Hamilton equations can be identified with the Principle of Least Action; which may be briefly and inadequately expressed as saying that the product of energy and time tends to a minimum, or at least to a stationary state, so that the path is determined by the vanishing of the variation of a certain integral expressing this product, called "Action". And this Principle of Least Action has been applied both in optics and in mechanics. In optics it was known as Fermat's theorem. Schrödinger recognised that the same principles served for both optics and mechanics; the very same equation might be interpreted optically, or might be interpreted mechanically. To give a rough illustration of this:

(1) In a wave equation the wave velocity v enters thus,

$$\sin p\left(t-\frac{x}{v}\right);$$

(2) in the Lorentz transformation an oscillation, to an observer travelling with relative speed u, is

$$\sin p\left(t-\frac{ux}{c^2}\right);$$

(3) so if $v=\dfrac{c^2}{u}$, the two expressions have similar form, in fact are identical. The optical velocity

v and the mechanical, or observer's relative, velocity *u* are reciprocally related.

The stormy or perturbed areas, characteristic of group waves, were thus found (when reasonably small) to move about under precisely the same laws that govern the motions of particles in ordinary mechanics. The frequency of the waves (their rate of vibration) appeals to us as the energy of an equivalent particle; and if acted on by other bodies, that is if they are immersed in a field of force, they acquire another peculiarity which may be called a potential frequency, dependent on the field in which they are, and representing their potential energy in that field. So this highly important part of Schrödinger's theory may be summed up, as it is summed up by Eddington, thus:

> "The equations for the motion of a wave group with given frequency and potential frequency are the same as the classical equations of motion of a particle with the corresponding energy and potential energy."

Now this is a tremendous generalisation, and no wonder it has been received with enthusiasm. What it all means we have still to find out, but it evidently means something important. Everyone is ready to admit that waves must be an etheric or spatial phenomenon; and now it is found that the right sort of waves are just what have appealed to our senses as matter, and what physicists and engineers have dealt with from time immemorial. The physics of group waves has risen to high importance.

Connexion between Form-Waves and Group Waves

But group waves are, after all, a resultant outcome of superposed constituent waves. They are a combination phenomenon, an organisation, as it were; they are a localisation of other waves much smaller than themselves, which are travelling at some speed dependent on their wavelength. The group waves have another speed, differing therefrom by an amount exactly determined by this dependence. This subject is so definite and simple, and so instructive as possibly illuminating the connexion between ether and matter, or ultimately perhaps illustrating one phase of the connexion between mind and matter, that a little elementary algebra may be permitted, here, to indicate the precision of our knowledge about the purely physical side of it.

What is called the wave "frequency" means their recurrence in time, their number per second. It is the reciprocal of their time period, or

$$\nu = \frac{1}{T}$$

What is called the "wave number" means their recurrence in space, the number per centimetre or per yard. It is the reciprocal of the wave-length, or

$$N = \frac{1}{\lambda}$$

The wave velocity, v, may be represented either as the ratio of λ to T, or as the ratio of ν to N. Let us say,

$$v = \frac{\nu}{N}$$

When two waves of slightly different frequency, ν and $\nu - d\nu$, are superposed, the resulting amplitude will vary periodically from double value to zero, and thus advancing groups or group waves will be formed; and the speed with which they travel will depend on the difference of ν and the difference of N (cf. p. 120). Rayleigh's quite general law for the speed of group waves is, in its simplest form,

$$u = \frac{d\nu}{dN}$$

Whether the velocities here denoted by u and v differ, and how much they differ, depends entirely on how much the main wave speed v depends on wave-length. The general law, which differs only algebraically (that is in nothing but form) from what we have already written in the above two equations, is that the velocity of group waves is

$$u = v - \lambda \frac{dv}{d\lambda}$$

This shows that if v is independent of λ, then v is equal to u. If v increases with λ, v is greater than u. If v decreases with λ, then u is greater than v.

Special Instances

In all this, so far, there is no reference to the ether. The theory applies to all kinds of waves. For instance, those large elevations on the surface of the sea travel at a rate varying as the square root of the wave-length,

$$v \propto \lambda^{\frac{1}{2}}$$

In that case $$u = \frac{1}{2} v;$$

so the groups travel half as fast as the individual waves.

For minute ripples on a tumbler or a teacup the speed varies inversely as the square root of the wave-length,

$$v \propto \lambda^{-\frac{1}{2}};$$

and in that case $$u = \frac{3}{2} v,$$

so the groups travel faster than the ripples. Or, in general, if

$$v \propto \lambda^n$$

$$u = (1 - n)v$$

It is all amusingly simple, once the theory has been given.

In sound and light v is really independent of λ, or $n = 0$. And so in those important subjects $u = v$. That is why group waves have not been much attended to in acoustics and optics. Every size of wave there goes at the same rate—a rate depending only on the medium.

Connexion with the Velocity c in Ether

So far we have been dealing with the subject generally, and in terms of material media, but when we come to the ether, with all its remarkable properties, we can proceed further. It is characterised by a constant velocity c, and in the simple case of light we know that $u = v = c$. That is true for the innate electrical properties of free ether, the properties with which we are most familiar. It is true except in so far as it is interfered with by the presence of matter or of free electric charges. Matter reduces the speed of wave transmission: conversely a number of free electric charges increases it.

For the ether has other functions to fulfil; it

is interfered with, and as it were "loaded", by matter; so its wave-transmission speed varies accordingly, and is not necessarily the same as its constitutional velocity *c*. Indeed we need not assume that there are no waves which can travel faster than *c*. In electrified space it turns out that waves can travel quicker than light; and the medium is then said to be super-dispersive.

But still *c* appears dominant. Why? What we know chiefly about the ether are its electric and magnetic qualities. Everything that we are accustomed to depends on them. They can be expressed or symbolised by Faraday's electric capacity-constant K and Kelvin's magnetic permeability-constant μ; which are *analogous* to an elastic- and an inertia-like quantity respectively. (Neither Faraday nor Kelvin used these expressions as anything more than the numbers relative to different materials, but they are essentially unknown physical constants of the ether, whether modified by matter or not. When used relatively they are only numerical ratios, K_1/K and μ_1/μ.) Clerk Maxwell showed that light was not only an optical but also an electromagnetic phenomenon, and that its speed depended on both these etheric constants, being in fact the reciprocal of their geometric mean, or

$$\frac{1}{\sqrt{K\mu}} = \text{velocity of light.}$$

This is the velocity that for free ether we call c, and it is a consequence of its electromagnetic properties. In free ether every length of wave travels with this velocity. Inside transparent matter the effective wave velocity is diminished, while inside an electric region it is increased. Everywhere that constant etheric velocity c is dominant. All other speeds are connected with it in some way.

We now begin to find or suspect that the ether has many other properties. Gravitation, for instance, does not seem to be electrical; and there may be others. There may be vibrations of various frequencies travelling at a higher speed. That speed is assumed to be dependent on wave-length, probably by some such relation as

$$v^2 = c^2 + b^2\lambda^2$$

for that is the law of dispersion in a known kind of super-dispersive medium, such as is constituted of electrified ether, or ether in the immediate neighbourhood of an electric charge; yes, even in the immediate neighbourhood of one electron. As to the other speed u, the velocity of group waves, it is always dependent on the velocity v, as indicated above in the general theory before "special instances", namely by the relation

$$u = v - \lambda\frac{dv}{d\lambda}$$

Combining these two last equations together, we get by simple algebra the profoundly interesting result that u and v (so determined), though they can vary enormously, are interlocked by the unalterable value of c, which is the geometric mean between them. The product of u and v multiplied together would in this kind of dispersion be constant and equal to c^2. So that whenever u decreases, v increases; and if u is ever zero, v must be infinite.

Now v in a given medium is not under our control; it has to adjust itself automatically. But u is sometimes under our control, even as far as matter is; we can deal with some group waves, since they appeal to us as matter particles. We may be puzzled to measure u, the speed of their or our own travel through the ether; we know it cannot exceed c, but we know no reason why in exceptional cases it should not reduce to nearly 0.

(Perhaps it can hardly be said that we are *sure* that the dispersion law

$$v^2 = c^2 + b^2\lambda^2$$

holds for the minute form-waves we have been considering; they have not yet entered physics in their own right; but for want of better knowledge we are assuming this as a reasonable law. If that is not true, then it may not be true either that

$$uv = c^2$$

There is sure to be some law connecting u and v, and this one is the most probable, but whether c is involved in precisely this way may be left at present moderately uncertain.

If, however, this simple connexion-law holds good, then the above law of dispersion follows as an inevitable consequence: for algebraically no other law of dispersion is compatible with that particular connexion between u and v.)

It is tempting to add, what Sir J. J. Thomson has pointed out, that the coefficient b in the above super-dispersion formula represents the smallest frequency which these form-waves can have in the ether. This minimum is a very high frequency, beyond that of X-rays, and there is nothing to prevent the existence of still higher frequencies. Nor do I see any physical reason for denying a similar sort of theory to very much longer waves; but we are dealing now with the small ones. The wave-length associated with a particle flying at the speed u we know both by theory and experiment to be given by

$$u\lambda = \frac{c^2}{b}\sqrt{\left(1 - \frac{u^2}{c^2}\right)}$$

Abbreviating the so-called FitzGerald contraction factor $\sqrt{\left(1 - \frac{u^2}{c^2}\right)}$ into $1/\beta$ as usual, the group-wave frequency follows algebraically from this equation. Written most simply, this equation is merely $\beta bu\lambda = c^2$, and is depicted in the figure on page 142; where, by the way, β is incidentally represented by the ratio $AB : BD$, so that this ratio, inverted, corresponds to the FitzGerald contraction. The group-wave frequency follows at once as

$$\nu = \beta b$$

This frequency will have a value directly proportional to the energy of the particle, for it is generally recognised now (see p. 125) that

$$mu\lambda = h, \quad \text{or} \quad h\nu = mu v = mc^2$$

and that $m = \beta m_0$; this being the mass of a moving electron, or of a particle on the electric theory of matter.

So that
$$b = \frac{m_0 c^2}{h}$$

Reckoning b thus, it comes out

$$1{\cdot}2\dot{4} \times 10^{20} \text{ vibrations per second}$$

Whereas Professor G. P. Thomson's experimental measurements of the diffraction patterns or rings, obtained on his photographic plates from the impact of high-speed electrons, make the minimum frequency

$$1{\cdot}08 \times 10^{20} \text{ vibrations per second,}$$

which, though not exactly the same as the above calculated *b*, is of the same order of magnitude.

It may be added that when the speed of a particle is insufficient for waves to accompany it, it is still able to interact with an electrostatic field; and this represents the sluggish deposit or residue of the vibrations. As the particle approaches rest, the magnetic part of its energy vanishes; its energy is then wholly electrical.

Sir J. J. Thomson has shown that in a superdispersive medium the magnetic energy need not equal the electric energy of the waves: "the difference between the electrostatic and magnetic energy is on the average equal to the kinetic energy possessed by the charges which give rise to the dispersive quality."

Lord Rayleigh first showed that the rate at which waves in general transmit energy across unit breadth depends not upon the speed of the waves themselves but upon their group velocity, which in the case of water waves is only half as great as *v*. A general law is that the amount of energy transmitted in unit time is *u* times the average amount of energy in unit length of the system of waves.

Locomotion and Energy

It will readily be perceived that the only reason for treating a group wave as due to the superposition of *two* similar waves of slightly different pitch is because that example is so simple. It is easier to add together two sines than it is to add a multitude, but in nature we must assume that, if there are ether wave forms at all, there must be an immense number of them travelling with all sorts of speeds between *c* and infinity, and with all sorts of wave-lengths. The ether I regard as the seat of tremendous rotational energy, but none of this need travel with the waves; locomotion is not a feature in free ether, locomotion is only introduced when the ether is modified into particles, or, what as we now think is much the same, into group waves. A group wave, or region of special disturbance involving energy, can travel at any speed from 0 to *c*, and that is how locomotion comes in.

Whether there are any quantum-like discontinuity conditions among the wave-lengths possible in free ether, or whether the variation in wave-length is continuous, I do not know. In neither case will there be transmission of energy except in isolated or specialised regions in which group waves have been formed, or (more generally) in which particles such as electrons and protons, whatever their nature, have crystallised out of the unmodified ether and taken on the power of locomotive transmission of energy and substance. It may be that only thus is what we call "substance" born, and that inertia then comes into being. Though, as I think that energy must be there in stationary rotational form all the time, I am inclined to regard the whole of the ether as in some sense substantial, and as endowed with the property which, when displayed, we call inertia. Weight, however, or gravitative attraction, I would associate only with the modified regions that are susceptible of approach and recession or other varieties of locomotion. Some sort of appreciable "identity" seems necessary in order to make gravitation an effective reality. At the same time the whole ether is subject to immense intrinsic pressure, and the gravitational influence displayed by localised particles may be, and probably is, a residual effect of that tremendous pressure.

CHAPTER VI

GUIDANCE AND CONTROL

Guidance by Wave-Forms without Energy

I have already answered the question how waves of any kind can travel faster than light. Such waves seem to be mere forms, they are not open to investigation, they make no appeal to our instruments or our senses, no signal can be sent by them, for they convey no energy, the energy is all associated with the group waves. When those group waves travel slowly, the constituent waves are quick. The group waves follow a path determined by the constituent waves, which therefore act as a guiding and directing agent, elusive in itself, but important as exercising control. The present idea is that certain etheric group waves constitute matter, and that this is a form of energy capable of being guided by something other than energy, something which acts as a guiding or directing principle. The guiding waves and the energy waves are interlocked or interacting, and yet are distinguishable from each other.

The whole thing is puzzling, because we haven't got to the bottom of it, but the indications of theory about velocities can at least be represented by a simple diagram. Draw a semicircle, let the horizontal base represent *c*, and draw a slant line from one end of the diameter to meet a vertical from the other end at a height proportional to wave-length. The length of that slant line is v, the wave velocity; and the portion of it intercepted by the circle is u, the group

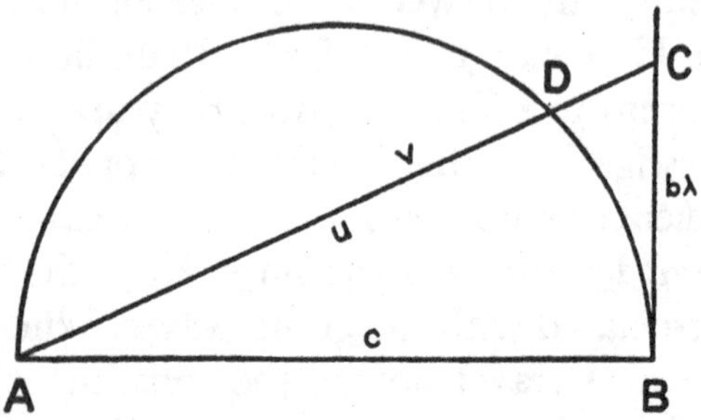

If $AB = c$ and $BC = b\lambda$,
then $AC = v$, the wave velocity or speed of the guiding waves, quicker than light;
and $AD = u$, the group velocity or speed of the energy waves, slower than light;

the curve being a semicircle on AB.

It is plain that $u = c \cos \theta$
while $b\lambda = c \tan \theta$;
so $ub\lambda = c\sqrt{c^2 - u^2}$
or $AD . BC = AB . BD$
as virtually said above (see p. 136).

velocity. The amount of slant may be anything, according to circumstances, from 0° to 90°. At 90° the matter, or the group wave, or the energy, would be at rest—whatever that means—and the speed of the constituent waves would be infinite. An approach to the other extreme case, with the amount of slant nearly zero, is what we encounter in the Heaviside layer through which wireless waves are travelling; the signals, which correspond to locomotive energy, travel with the group wave velocity, nearly equal to but always a little less than that of light. So then the three velocities, u, v, and c, are very nearly equal; but v is a little bigger than c and u a little less. The signals are guided by, or follow the path of, the component waves, of which the upper portions travel faster than the lower—because the upper air is more ionised than the lower—so that they can curve round the earth and reach the Antipodes, as explained by both Eccles and Larmor. The longitudinal contraction which matter appears to undergo whenever it travels through space with velocity u is represented by shrinking the length AB till it equals the length BD. In other words, the trigonometrical sine of the angle of slant represents the FitzGerald contraction.

Psychical Speculation on this Basis

I must not dwell on these physical considerations any more, for I want now to ask a question which I imagine philosophers and psychologists of the future will have to tackle, however remote from their present ideas such a conception may be.

Hitherto we have had no permanent physical

basis for the familiar entities called life and mind, which I regard as essentially one and the same thing in different stages of development. The group waves have been found to subserve the functions of matter. What use is to be made of the constituent waves? They are not of a totally different character from the group waves, they interact with them, they guide them, indeed they make them possible; they are the substratum of everything, but they make no sort of experimental appeal: our observations are limited to the group. With the group only can we deal experimentally or observationally. Yet surely the component waves must have some function. We may be using them all the time without knowing it. It is found that those constituent waves are a guiding and directing influence on the particle; it is those which determine the path of the energy. They constitute a guiding and directing agency; they have no energy of their own, but they achieve results which would not otherwise be achieved.

It must be apparent that these are the kind of assertions which we have been constantly making about life. Life is a guiding and directing principle. It controls matter and energy, with a certain element of spontaneity, and yet it differs from both. What we have wanted is a physical basis for such activity. The presumptuous and hypothetical suggestion which I want to risk is that these constituent waves of excessively high frequency, far higher than anything we have yet apprehended, may be the physical basis of life and mind. The ether is hypothetically full of them, their nature is unknown, and yet they are responsible for everything that exists. I suggest that they form the physical basis, though not in the least a material basis, for an idealistic interpretation of the universe, in which life and mind are supreme. They are a concomitant of the slow-moving group waves which constitute matter, and they can travel at a nearly infinite pace; one is tempted to say like thought.

Bergson held that the Ultimate Reality, for purposes of evolution and development, had differentiated itself into two interacting portions; one of which we knew primarily by instinctive apprehension, the other of subordinate and, so to speak, lower grade, which appealed to us through our senses. He suggests that this "matter" portion was degraded in order to get the variety and the interest and the possibilities associated with two interacting things. By degrading a portion of itself, the original supreme entity could thus rise on stepping-stones of its dead self to higher things. A rise in the scale of values, a development of existence, might thus be produced which could be attained in no other way. Monotony of existence would cease, struggle and effort would begin, if life for a time could leave the constituent space-waves and enter the groups. For the groups have an identity of their own; and in that way life might become individualised. It would thus acquire new powers, encounter new difficulties, attain new experiences.

The interaction of ether and matter is studied under radiation, which is an electrical process of interchanging energy; and the result of the interaction is either the emission or the absorption of a quantum of ether waves travelling with the velocity c. These are familiar in physics, and are the basis of optics. Certain group waves travelling with the velocity u are familiar in mechanics, and constitute the basis of matter. What about the waves with velocity v?

My expectation is that some day psychologists will find that these at present dimly inferred and barely apprehended high-frequency periodicities may be the instrument utilised as the physical basis for that other highly familiar phenomenon in the universe, our own life and mind; the general nature of which, apart from any physical vehicle, we are acquainted with by first-hand experience.

How spontaneity and free will and planning and purpose can thus be given a physical basis I do not fully see; but they must have a physical basis, else they could not interact with matter. The problem before us is only beginning; but it is in this way that I would seek for a solution of the problem of spontaneity or freedom of the will. Not in Eddington's way, by a postulated defect in the law of causality when dealing with small individual particles. Break down causality, and we are left with chance. That is wholly unsatisfactory. It may be true that the jumps of electrons in the atom cannot be predicted, they often seem to occur by chance. But not in that way would I aim at freedom, for chance is no solution. The laws of Probability only apply to a multitude, not to an individual. There must be a cause for each of those jumps. And the cause is already being sought. Some physicists attribute radio-activity to the impact of extra high-frequency radiation. The recently discovered cosmic rays are already thought to have an influence on biological phenomena, as they certainly have an influence on atomic processes. High-frequency vibrations, not generated on the earth, nor in the solar system, are now at last being discovered and attended to.

So now that we have found out that matter is a localised peculiarity in the midst of an ocean of excessively rapid periodic influences, there is plenty of scope for the idea that the seat of our intuitive apprehensions is to be sought in what we have called empty space; that whatever the ultimate nature of those influences may be, space is their habitat. We may presume that those periodicities are truly animated, and that they constitute the agencies or instruments through which we operate in the material sphere. The whole universe seems to be an animated structure far beyond our present apprehension; we surmise that intelligent beings are making use of its constitutional periodicity; and it is turning out that upon some etheric periodicity or wave structure we ourselves and all animated beings are wholly dependent for any display of our vital and mental activities.

CHAPTER VII

INFLUENCE OF ORGANISATION

We may try to carry the subject a little further. But first I would remind you once more, as a brief summary of what mathematical physicists have taught us, that although the great bulk of ether is for most practical purposes a homogeneous medium in which waves of every length travel at the same rate, yet any region of it which contains electric particles will be what is called a super-dispersive medium, in which waves may travel at any pace according to their wave-length. There may be group waves in a medium of any kind; but the peculiarity of a super-dispersive ether is that its group waves travel slower than light, while the waves which constitute and guide them travel quicker. The velocity of light is the geometric mean between the two other velocities of transmission, that of the group wave and that of the constituent or guiding wave. Ether waves appear to be able to prepare an easy path and thus guide flying electrons in their course. The possible implications of this we have already attended to.

We are also told by Sir James Jeans that an electrified particle, say an electron, so long as it is free and not associated with any other particle of opposite sign, is incompetent either to emit or to absorb light. It can be pushed along, or rather guided, by waves; but it has no means of absorbing them. To be active as a radiator or absorber it must be associated with another charge of opposite sign, that is, it must be a constituent of an atom. An element of organisation has then been added. In any organised system of electrons, such as an atom, the individuals have acquired a new property due to the organisation—perhaps many new properties, but at least they enter into a new relation with the waves in space, they can now absorb or emit quanta. An incipient element of energy-exchange and effective activity has been introduced. Somewhat as the energy of gunpowder is more effectively utilised and absorbed by a bullet when it is loaded into a gun; for then the bullet is fired off in one direction while the gun is fired in the other, with equal opposite momenta though with very different energies. Omit the gun, and the bullet is not projected. Organisation into an atom has improved the conditions for energy interchange between radiation and electrons, i.e. between ether and matter.

This enhanced power of an electron, directly it is incorporated or organised into a constituent part of an atom, is perhaps the most incipient example of what has been called "emergent evolution", an organisation acquiring functions not possessed by its constituent elements while separate. Common experience illustrates this notable outcome of organisation further. It is proverbial that neither an atom of hydrogen nor an atom of oxygen has any of the properties of water: the separate atoms acquire those properties by combination. Hence a molecule has properties denied to an atom: and so also an organised assemblage of molecules is found to have properties which the separate molecules do not possess. An organic compound is an assemblage of a large number of molecules; and if the molecular compounds are sufficiently complicated and sufficiently numerous, so as to be organised into a portion of protoplasm, then a further property makes its appearance. A protoplasmic cell has the chance or the opportunity or the potentiality of being endowed with vitality. It need not be living, but it may be. It has become able to receive and store the unknown spontaneous guiding function that we call life. It may exhibit the properties of a living cell; for thus has life found it possible to enter matter and become incarnate. And when these living cells have aggregated still further, and become organised into a portion of brain, then a still higher function is able to manifest itself; the brain cells can then act as the material vehicle for thought and mind. It must not be supposed that the fact of material organisation has brought these higher entities into existence; *the function of organisation is not generative but demonstrative*; it enables or induces sensory apprehension and consequent display; it makes the latent conspicuous. It embodies the abstract. The main realities in the universe lie in the region of the ideal, the supersensual, not in the material vehicle which partially displays or manifests them.

Without attempting any explanation, I claim that these functional adaptations of complexity are facts of common experience. We have learnt them by observation of the things around us. The explanation is still to seek. But it is instructive to realise, and it has often been emphasised, not only by Professor Lloyd Morgan, but also in other forms by General Smuts, and by Professor Whitehead and Dr. J. S. Haldane, among others, that an organism as a whole has properties which do not belong to its constituent parts; and that we ourselves, being such organisms, are conscious of what those properties feel like, and are aware of what they can do. We cannot reason it out but we know it by direct experience. It is surely of interest to find, what seems to be indicated now by the higher physics, that this gradual increase of function with organisation can be traced down in regular succession till we reach the simplest units of all, even down to the electron and the proton, which acquire properties when combined which they did not possess when separate. Another illustration from the widely prevalent phenomenon of sex differentiation is obvious.

Emergence

Now the question forces itself upon us, how can these new properties emerge if they are non-existent in the elements? I would not venture to say that they are not existent: I would only say that they are not displayed. Essential realities must surely pre-exist somehow, somewhere, even though imperceptibly. Emergence means demonstration or display; it means recognition of what may have been latent before. We find that animation is only displayed by material aggregates which have developed into organisms. Analysing the organisms, we lose the animation; just as if you microscopically examine a picture you lose its meaning. We analyse the organs of mental activity into protoplasm, we resolve that into organic molecules, the molecules into atoms, the atoms into electrons, and the electrons more or less into group waves, while the waves themselves are periodic disturbances in an unknown omnipresent medium with physical properties, which is sometimes identified with space.

I feel compelled to assume that the latent possibilities of animation are already existent in space, but that they cannot come to fruition or display themselves, at any rate to us in our present condition, unless they interact with matter. And it appears that physics is now helping us to trace the gradual possibilities of interaction between space and matter. What we have recently found, or are tending to find, in space, are certain excessively rapid periodic disturbances; though the word "disturbance" is not a good one, since it suggests irregularity, which is not intended. It seems hardly legitimate to call them periodic *motions*, though for want of a better idea that is how we are inclined to think of them; we are only entitled to say a periodic "something", constituting an excessively fine-grained kinetic structure, of nature unknown, which here and there has aggregated itself into group waves. These either constitute or are associated with what we have long called an electric charge, which thus in wave form for the first time enters the field of physics, and by its further interaction with other charges, builds them up into atoms and molecules and all the other structures that we are acquainted with, and enables them to move in accordance with certain laws of motion which we have more or less thoroughly detected.

Our sensory apprehension is limited to matter, and we know that either directly or indirectly some of that matter is animated; for organised aggregates of it appeal to our senses. It may be that there are other organisms in space, more immediately in touch with its constituent periodicities, and not needing any aggregation into material forms. These therefore are fairly inaccessible to us. The life and mind which in such a medium are to us latent, and seem practically non-existent, may be, I suggest, only non-displayed because of our own limitations. All the possibilities that we have perceived must be there, in some sense, from the beginning; though all that we assuredly observe or infer is, first, the material vehicle which has demonstrated those properties, and then the more elementary forms which conceal them.

But beyond all these we are beginning to detect a physical or etheric basic organisation, and to suspect that it is utilised by, though it does not directly demonstrate, still higher unknown attributes, of which we, through our higher organisation, find in our own consciousness some surprising glimpses. We do not understand those higher properties any the more for our analysis: they are recognisable only by direct apprehension. But the research into the physical vehicle, which presumably all the time they must be utilising, is a permissible one. Part of the analysis has been conducted by the legitimate method of physics, and part by the illegitimate speculative method which must be treated with suspicion and caution and constant emendation as going "beyond physics".

APPENDIX 1

ON BIO-PHYSICS

IT is rash for a physicist to trespass on biological territory; but of late there has been an indication that barriers are beginning to break down, in a way rather suggested by the names Bio-Chemistry and Bio-Physics; and a new technique has been evolved by the biologists themselves, which is of great promise.

In comparatively recent years the discovery has been made that living tissues can be nourished and kept alive as specimens in glass vessels, and can be so arranged that the processes of growth, cell-division, locomotion, reproduction, phagocytosis, and metabolism generally, can be watched under varying conditions, and be subjected to physical and chemical stimuli under the microscope. Apart from what is being done in America and Europe, these operations are being conducted in several places in this country, and perhaps chiefly in the Strangeways Research Laboratory at Cambridge, opened and initially endowed in 1912 by the enthusiasm of the Cambridge University Lecturer in Special Pathology, the late T. S. P. Strangeways, and a few medical friends. In May 1929 Dr. Canti, of St. Bartholomew's Hospital, exhibited to the Royal Society a kinematograph film on which successive stages of several microscopic cellular processes had been photographed by him, from cultures grown at this laboratory, so that their projections on a screen could be run through rapidly and the cell-motions easily followed by ordinary observation; though to understand what was happening in detail to the living cells under the microscope a good deal of special knowledge and previous experience must naturally be required.

Stimulated by this demonstration, I paid a visit to the Strangeways Laboratory, and renewed my acquaintance with the subject, which had begun for me at the Rockefeller Institute, New York, in 1920. There I had seen some tissues that had been kept alive for months or even years, by what may be called the simple plan of supplying them with nutriment, removing any poisonous secretions, or in other words keeping them clean, and by excluding the microbes whose entry would result in decomposition and putrefaction. Since then the technique has gradually improved, and presumably has become more skilled and elaborate; so that interesting observations can be made both on healthy and on diseased cells, and so that the effect of drugs, and physical stimuli such as radiation, can be directly observed in microscopic detail, without the complications inevitable when the tissues still form part of a living animal. In general terms, it may be said that vital operations can now be watched, and the phenomena of life studied, in their simplest and most rudimentary form.

On the 8th January, 1930, Dr. Honor B. Fell, Head of the Strangeways Laboratory, speaking at the B.B.C. in London, broadcast an elementary idea of the technique involved, and thereby called the attention of a wide public to what has been going on behind the scenes for a considerable time.

A discovery of this kind can only be in its infancy; it is bound to have extensive developments, and cannot fail to have ultimate application to scientific medicine. Already the way in which the white corpuscles or phagocytes swallow or engulf diseased cells can be seen in operation—an active cell, with obvious exertion, draws itself outside an enemy cell as big as itself, and thereby renders it harmless—an interesting thing to watch. I understand also that the effect of high-frequency radiation on cancerous tissue can be examined in detail. Hence it would appear that no longer need those engaged in physiological and pathological researches be working in the dark. They can actually see what is happening, under refined and simplified conditions; and it seems difficult to exaggerate the importance of such a discovery. Physicists, chemists, and biologists will be brought together, younger workers will be trained, medical practitioners can co-operate; and the ultimate outcome must be something that no one can yet foresee.

No doubt funds must be forthcoming, the work must

grow and extend, more laboratories must be established, until gradually the operations of life can be studied with the same scrutiny and understanding that has already been applied to the behaviour of inorganic molecules and atoms. The overlapping borderland of sciences has often proved fruitful; and now that physics, chemistry, and biology are being brought together, and a method devised whereby minute portions of living tissue can be subjected to all manner of influences—now that the behaviour of living cells can be watched under specially arranged conditions—no limit can be set to the possibilities of further discovery. And surely it is of a kind which promises to have a direct and beneficent application to the treatment of disease and to human needs generally.

Hitherto the effect of chemical and physical stimuli, and the results of inoculation or the action of one group of cells on others, have had to be conducted under difficulties, by watching the comfort or discomfort of live animals subjected to treatment. This latter method of diagnosis, effective but rather indirect, has aroused a good deal of opposition from kind-hearted though often misinformed people, who have qualms about the legitimacy of such procedure; they doubt whether the information gained, valuable as it is, and inevitable if no better plan can be evolved, is sufficient to justify the artificial subjection of living creatures to pain and discomfort. It may be hoped that the activities of kindly people, which have hitherto been exercised in a negative direction, will now take a positive turn; and that they will ease their consciences and use their resources to encourage the new methods of investigation, and to promote the study of vital activities and cellular changes as exhibited by microscopic tissue-cultures, apart from any sensitive living creature. Experiments are thereby partially freed from the complications of a complete organism, and are conducted under circumstances on which a growing number of observations can be made in the most direct manner, without pain or discomfort of any kind.

Those also who are interested in the nature of life, and what the vital behaviour of matter really means, will surely here find an opportunity for studying life in its simplest form. It is extraordinary, for instance, to see the process of bone formation beginning and continuing in a microscopic specimen of soft tissue, such as that of an embryo jaw, or a limb, and to realise that the formation of various organs depends on a local system of control, acting without any necessary connexion with a central nervous system and apart from the rest of a complex organism. Twenty years ago no one could have supposed that an embryo heart, removed from a partially incubated egg and cut up into small portions, could continue to beat for a considerable time: or that a fragment of tissue removed from the dead body of an animal could be revived and preserved in a living condition for an indefinite period. Yet these things can now be done, some of them easily, others with care and proper precautions, by those who have acquired the necessary technique; and if the present generation facilitate the work, the ultimate outcome can safely be left to posterity.

So far there is here no question of generating life *de novo*. There must be a nucleus of antecedent life; but, given that, a living tissue not only continues to exist, but increases and multiplies and goes through all the operations which hitherto have been studied only in complete organisms. The mystery of life remains, but its behaviour has been simplified.

I may remind readers that similar advances have been made in other branches of knowledge: At one time electricity was mysterious and its nature unknown; the discovery of the electron did not solve the mystery, but reduced it to a simple elementary form, and revolutionised the treatment of the whole subject. The resolution of magnetism into molecular electric currents was a great step in advance. The discovery that light was an electromagnetic oscillation was an illuminating theory, and has resulted in remarkable applications. A discovery of the intimate structure of the atom, although much remains unexplained, was a step of fundamental importance. It may be that a correspondingly momentous step is now beginning to be taken towards an understanding of the interaction between life and matter.

APPENDIX II

ON THE NATURE OF PHYSICAL EXPLANATION AND ON THE VARYING FOUNDATIONS OF PHILOSOPHICAL PHYSICS

UNTIL the twentieth century the simplest and most fundamental phenomena to which it was hoped to reduce everything else were those treated of in the mechanics founded by Galileo and developed by Newton. This mechanical or engineering foundation embodied axiomatic conceptions based on universal experience; and nothing simpler or more fundamental could well be imagined. Everything depended on the three fundamental entities, length, time, and mass; or, as they might be expressed, space, time, and matter. The sensations underlying and justifying these abstractions were contributed by our muscles and nerves, as a sense of free motion, of enduring motion, and of resisted motion. Or we might summarise our direct apprehensions as speed, fatigue, and force. Upon these experiences our apprehension and formulation of three dimensions of space, the steady flow of time, and the permanent inertia of matter, were securely founded.

Derived terms speedily followed:—

The ratio of speed to time was acceleration.

Power was the product of force and speed.

Inertia was the ratio of force to acceleration, and was called mass.

Expended energy was the product of power and time.

The product of mass and speed was momentum, the result of a force enduring in time.

The product of momentum and speed was kinetic energy, the resulting product of force and distance, or range.

Angular or rotational momentum could be expressed as the product of energy and time, or as the ratio of moment of inertia to a time-period.

Mass near the earth had weight, and this formed a convenient statical measure of it.

These relations remain valid, though they take on different dresses as time progresses and fashions change.

There were things that would not fit satisfactorily into this mechanical scheme: and many rebelled against its being regarded as a comprehensive outline of the universe. Not only did vital and mental experiences seem alien, and insusceptible of a mechanical explanation: even the ordinary phenomenon of "light" was recalcitrant. Heroic efforts were made to develop a mechanical theory of light, but the result in the long run was failure. Light was an affection of space, and what space was like no one knew; for it made no appeal to our senses, and was not included in the mechanical scheme. This defect was emphasised when new departments of physics arose. Electricity and magnetism were perceived to be also affections of space, and their nature, too, was unknown: only a generalised dynamics could apply to them. Gradually it was suspected that electric charge might be more fundamental than matter: and instead of explaining electricity in terms of ordinary experience, the operation might have to be reversed. The properties of space could not be explained through an equipment of mechanical principles developed in association with matter alone.

So a new doctrine sprang up, that all our direct experience was relative, and that we must seek for absolutes in the unknown properties of space. Even kinetic energy, expressed as $\frac{1}{2}mv^2$, was relative to some other body (usually the earth); for the only motion we could measure was motion relative to some material landmark. If we could detect motion through empty space that might be absolute. The only thing whose speed of travel through space could be ascertained was light; so that speed was taken as a first absolute. It is now conceived of as a constitutional velocity characteristic of space, which becomes known to us as the speed of transmission of the periodicities known as ether waves.

The energy of such waves was found to depend on their frequency or rate of vibration always. So the expression

$h\nu$ for radiation energy came into vogue, with h as an absolute unit of angular or rotational momentum. At the same time the individualised or localised ether energy associated with a piece of matter was mc^2; in other words, matter was recognised as a concealed form of energy, which only under exceptional (say stellar) conditions could be transmuted into other forms.

The electrical theory of matter had shown that in so far as a body was affected with a relative velocity v, its energy was increased by a factor $\beta = \frac{c}{\sqrt{(c^2 - v^2)}}$, so that its measured or relative mass was not constant but became $m = \beta m_0$.

The corresponding idea of mass, as the ratio of constitutional energy to c^2, could be applied also to radiation, which might thus have a sort of individuality conferred upon it or detected in it; so that $\frac{h\nu}{c^2}$ became called a photon, and was regarded as the unit of light.

The energy of an electric charge is the product of quantity and potential. An absolute unit of charge was discovered called e, so its mass was proportional to $\frac{e^2}{rc^2}$; and this was called an electron, on the supposition that it had a definite boundary of radius r.

Directly it was admitted that a photon could have mass, a revival of corpuscular light was inevitable; and as the photon was associated with a certain frequency corresponding to its energy, the idea of a vibrating particle or wavicle came to the front. Thus arose quantum mechanics, where the interchange of energy between ether and matter took place in countable units, governed by an absolute space-constant h. If $\frac{h}{2\pi}$ represents a unit rotational structure in the ether, circulating with the velocity c, I estimate the linear dimensions of that structure as $3{\cdot}8 \times 10^{-13}$ centim., which makes it look like an electron.

Associated with an electron it was found there would be a definite frequency, so that its energy could be expressed either as m_0c^2 or as $h\nu_0$. Moreover, as soon as a unit or indivisible quantum of light could be dealt with and defined, it was perceived that there was a great similarity between particles and waves. Every mechanical velocity u could be identified with a wave velocity $\frac{1}{v}$. For whereas the principle of Least Action in mechanics gave as the path of a particle between two points,

$$\int_A^B uds = \text{minimum}$$

Fermat's principle of Least Time in optics gave as the path of a ray of light,

$$\int_A^B \frac{ds}{v} = \text{minimum}$$

so that, if $u = 1/v$, the two expressions would be identical.

Thus began an association between mechanics and optics, which is essentially a diverting of attention from matter and concentrating it on space. Wave mechanics came into being, and the ether began to come into its own.

Group waves in ether are associated with wave forms in such a way that $uv = c^2$; group waves obey the laws of dynamics; so matter might consist of group waves and be guided by form waves. Form waves transmit no energy, but they exercise a guiding and directing influence.

Now opens up the possibility of extending physics, the physics of conjoint matter and space, till it covers the hitherto excluded phenomena of life. Life transmits no energy, but exerts a guiding and directing influence. In association with form waves it might do most of what is wanted. Gradually biology and psychology may be thus allied with physics.

APPENDIX III

SUGGESTED MECHANISM FOR GUIDANCE AND CONTROL

On page 38 and elsewhere I have drawn an analogy between magnetism and life. To effect a guiding influence on a material world, life must be able to deflect a material particle such as an electron. That a magnetic field can deflect the path of a stream of particles like cathode rays has long been known. That ether waves or wave forms, travelling faster than any particle, can likewise exert a guiding influence on the destination of the particle, by directing its course from point to point, has recently come to be known. The ether seems to be full of stationary periodicities, which, though they contain a vast store of rotational energy, behave something like Osborne Reynolds' detached pendulums, and do not transmit any; though by the sweep of a disturbance, which we might picturesquely think of as animation, they may constitute themselves into form-waves of any arbitrary length and corresponding speed, such as might suffice to form groups travelling with the speed of a particle, so as to guide it in a direction closely related to the normal of their wave front. If the particle were in motion the direction might differ from the wave-front-normal by an aberration angle; and there would be another definite angle appropriate to a magnetic field, or to some other modification of space.

Some kind of ether-modification we might not unreasonably regard as representing the physical side of animation, and as embodying the mechanism whereby life and mind are able to control the behaviour of matter without either adding to or subtracting from its energy. On this view, whatever energy corresponded to animation would be concealed, for energy-transmission would in general be zero, though wave forms could travel at any required speed. It would be greater than light when they were active. The guidance of a particle would be likened to group-waves guided by form-waves, and might correspond to a slowing down of the wave forms; so that for the special case of a particle moving nearly with the speed of light, the wave forms would slow down to near that same speed. But such a particle would be hard to guide, or would be characterised by great inertia, and an actual beam of light would probably not be subservient to guidance at all, except by material means. Thus light is a limiting case of matter. Matter is able to influence and direct light (e.g. by mirror or lens), just as matter is able to deflect other portions of matter, as illustrated by a cricket bat or a planet. On the wave theory of matter, deflection is hypothetically effected on group waves travelling at any speed less than *c*.

Every determination of the velocity of light is made by observing an identifying group-feature purposely imposed upon an otherwise continuous beam. It is by the fluctuations of amplitude imposed microphonically on a carrier wave that we are able to produce those group waves that can subsequently be translated or interpreted as speech and music. Such group waves bear the impress of mind. May we not say that the orderly movements of particles in growing tissue, and at a higher level in brain cells and nerve fibres, demonstrate the directing and controlling influence of animation, or of life and mind.

Orthodox Citations

In support or further illustration of these views I shall quote sentences from two physicists, one an ultra-modern Professor, one an Ancient Master, neither of whom was thinking in my direction.

From a Friday Evening Discourse at the Royal Institution by G. P. Thomson on June 8, 1928:—

> Personally I see no necessity for there to be any vibration of a material or quasi-material object. The essence of a wave is the transference of a *state* according

to certain laws—for example, a "heat wave" crossing the Atlantic. The state may, or may not, be one of motion. On this view the function of the waves is to produce conditions at the electron which oblige the latter to move along the instantaneous position of the wave-normal (in the absence of a magnetic field), and along another, but determinate, direction in a magnetic field. But though it seems possible to regard the motion of the electron as determined by the wave in its immediate neighbourhood, it must never be forgotten that the direction of each part of a wave-front depends on the history of the whole wave-front. Any circumstance that influences one part of a wave influences the whole, and this is of the essence of wave motion. The electron is influenced by conditions at a distance, but indirectly through its waves. . . .

The easiest way of looking at the whole thing seems to be to regard the waves as an expression of the laws of motion. The uniform motion of Newton's first law is replaced by a simple plane wave, and so on. On this view, the electron remains the reality; and this seems right, for after all it is the electron as a particle which is actually detected in any conceivable experiment, and the waves come to bear the same sort of relation to it that Newton's or Einstein's law of gravitation bears to the planets which obey it.

Thus a modern physicist.

Now for an inspired utterance from the past. (Compare Query Seventeen in Book III of Newton's *Optics.*)

Those that are averse from assenting to any new discoveries but such as they can explain by an hypothesis, may for the present suppose, that as stones by falling upon water put the water into an undulatory motion, and all bodies by percussion excite vibrations in the Air; so the Rays of Light, excite vibrations in the refracting Medium or Substance . . . that the

> vibrations thus excited are propagated in the refracting or reflecting Medium or Substance, much after the manner that vibrations are propagated in the Air for causing Sound, and move faster than the rays so as to overtake them . . . and that every Ray is successively disposed to be easily reflected or easily transmitted, by every vibration which overtakes it. But whether this Hypothesis be true or false I do not here consider.

Newton thus emerges through the ivory gate. But, now that "rays" consist of particle-like quanta, his dream may perhaps without offence be paraphrased somewhat thus:—

We may suppose that as ripples are excited on water, and sound vibrations in air, by moving or impinging particles, so vibrations can be excited in the ether; the waves being transmitted by the properties of the medium faster than the particles, so as to overtake them; the path of a particle being determined by those overtaking vibrations.

APPENDIX IV

SOME EXPLANATORY NOTES ON TERMS AND IDEAS ABOUT WHICH READERS HAVE EXPRESSED DIFFICULTY

Truth or Convenience

Partly perhaps from a sense of modesty, with the idea of curbing the extravagant pretentions to cover the whole field of knowledge, some scientific men have occasionally minimised the scope of their subject, and have said that science was not out for truth but for convenience of statement; that our laws of nature, for instance, constituted a summary or generalisation from observations; and that if they achieved statements of that kind, it was all that was to be expected of them.

There is a philosophic doctrine called Pragmatism which tests the truth of propositions by asking whether they "work" or not. If they make no difference one way or the other they are negligible; but in so far as they summarise the situation and express the facts, they are regarded as true. The law of Convenience is a further step in the same direction; and the implication is that a statement can be a convenient and handy shorthand summary without any more ambitious claim to be fundamentally true.

Many of our statistical laws really may be considered as of this kind. They deal with the average result of a huge number of special instances, which are too complicated to be followed in detail. The pressure of a gas, for instance, is due to the bombardment of millions of atoms having all manner of velocities, among which there is an average velocity which might be defined as that which would give the same pressure if every atom possessed it. The motion of the atoms, however, is perfectly irregular, and no one atom has the same velocity for a second together. And so in our statements the mean velocity is used as a matter of convenience, and the pressure of the gas is equal to one-third of its density multiplied by the square of this velocity, and is considered to be uniform. The number 3 that appears in the expression corresponds to the three dimensions of space.

When we are dealing with large surfaces the pressure of a gas or other fluid is uniform. A body the size of a cricket ball hung in stagnant air or water, though it is violently bombarded on all sides by the molecules, can be trusted to remain absolutely steady, because the pressure on every side is the same. There is no reason why it should be; it is simply a matter of averages. The molecules have every velocity, and consequently exert all kinds of pressure; but the number is so enormous that the average is uniform in every direction and for all time; so that the body submitted to the bombardment keeps quite still.

Yet if you reduce the size of the body to microscopic dimensions, it will not keep still; for every now and then the pressure on one side is for a moment greater than that on the other; and accordingly it is pushed hither and thither, and hardly ever stands still. The microscopic body becomes in fact one of the atoms itself, and shares in their movement, having as much energy as any one of them. But the energy of a moving body depends on its mass and on the square of the speed, and when the body's mass is very great, its speed is very slow. To see it move at all, the particle must be only such as is visible in a high-powered microscope. If we reduced its size still further, till it became no bigger than one of the molecules, it would be quite invisible, but it would then be flying about with the average speed of bullets, just as they are.

Hence the law of pressure, as ordinarily stated, only applies to bodies of reasonable size. A different law must be used when we are dealing with ultimate particles.

This is characteristic of physics. Its ordinary well-established laws fail when you come to deal with the ultimate particles of matter, and have to be replaced by others.

Hence, in a sense, such laws are conveniences rather than ultimate truths: for ultimately it is the particles themselves that are responsible for all the actions and all the activity: and to group them together into a solid body, and thus confer a sort of unity upon them, is in a sense artificial though convenient. But still the law of averages, Boyle's Law for instance, that the pressure of a gas varies inversely with the volume which unit mass occupies, is true whenever properly applied, remembering its limitations.

The statement that science is only aiming at convenience and has no concern with truth is not one to be allowed. Mach and Karl Pearson and Poincaré seemed to hold a doctrine of that kind, but in so far as they did they are not to be followed. The aim of science is absolute truth, and though it is difficult of attainment, nothing less should be admitted. Modern physics in especial is getting down as closely as possible to what is really happening to the ultimate particles of bodies and in the space immediately around them. Its task is difficult, because we have no direct experience of such minute and energetic happenings, and all our ordinary methods of investigation fail us. Hence for a time we take refuge in mathematical abstractions.

Pointer-Readings

Although Professor Eddington does not hold with the contention dealt with above, yet he admits another kind of limitation—a delimitation of frontiers which evades a conflict with Theology—by emphasising the fact that when we measure a quantity, part of the process consists in reading the divisions of a scale or the motion of a hand on a dial; in other words, he claims that when we specify a physical quantity we are really referring to the reading of some pointer or other. For instance, when we speak of the temperature of the Polar regions and say that it can go as low as $-60°$, we mean that a thermometer would read that value. And so also when we specify a wave-length of light: we mean that the interval between two adjacent crests would stretch thus far on a finely divided scale. And again the implication is that that sort of thing is all we are dealing with in science, that our equations refer to pointer-readings and to nothing else.

I suggest that that is confusing the measurement of a thing with the thing itself. When I speak of the cold of the Polar regions, or the conditions on the summit of Everest, I am not thinking of any thermometer or barometer, but of its effect on the nerves and on human life. I am as it were "feeling" the cold, or the lack of air, not measuring it. Similarly, if I think of the wave-length of sodium light, I am thinking of a multitude of things, but not of any scale or its readings; and it seems to me that excessive emphasis on the metrical character of science, represented by the term "pointer-readings," is another effort to minimise unduly the scope, the relevance, and importance, of a scientific statement.

Take temperature, for instance, which in a gas is a name for the average squared speed of the molecules. When Professor Eddington specifies the temperature of the interior of a star, he is thinking much more of the rate at which the atoms move about, and the consequences of that violent motion, than of any method of measuring temperature: in fact it is immeasurable by any thermometer, and cannot be read on any scale. It is a mental inference, and has a multitude of consequences not at all deducible from any pointer-readings. So when we say "the nineteenth century" we might mean that if we counted the revolutions of the earth round the sun there would have been time for 1,900 of them since the birth of Christ; but we don't really mean that; that is a metrical thing, not suggested by the term when we use it. So also when we say that a man is ninety years old. The phrase suggests many things, but not the fact that the earth has been ninety times round the sun since he was a baby. The Bank Rate again may be expressed and must be reckoned in terms of arithmetic, but the term to financiers means much more than that. The strength of an electric current can be read off on a galvanometer scale, but the reading needs a lot of other factors to make it intelligible; and anyhow, a pointer-reading is not the least like an electric current.

So in the laboratory we deal with a multitude of things about which we never read a scale except for some minor metrical purpose. In regarding the weather, we might specify the height of the barometer in figures; but weather means much more than that; and some of its features, for instance a thunderstorm, never get recorded metrically at all. But it is true that for its analysis and understanding certain measurements are necessary. And that fact is what gives plausibility to the contention that a study of pointer-readings constitutes science.

The term "science" ought to cover much more than its merely metrical aspects, and ought to extend to things which we do not know how to measure on a scale. In fact it was only in the last generation that electrical units were introduced. Measurement is artificial, yet even that involves much more than pointer-readings. Reading of an instrument may be what a foreman does in a factory, or a ship's captain does on the bridge; but things had to be organised thoroughly before that could be done, and even then more than that is required to keep machinery in order or to navigate a ship.

Relativity

A great many of our common ideas are relative, though we usually ignore the relativity aspect of them. For instance, there is nothing absolute about the motion of a railway train, but we state its speed as if there were, knowing all the time that it is relative to a moving earth. This affects the principle of the conservation of energy. The ordinary expression for it, $\frac{1}{2}mv^2$, which suffices for engineering operations, has reference only to the earth. Before Einstein, people knew this, but did not know the remedy; they did not imagine that a body had any absolute energy. Einstein set himself to discover absolutes, one test of which is that they would be the same to all observers, no matter what the observer might be doing.

But the only absolute velocity he was able to find was the velocity of light in free space, which according to his doctrine is constant and unalterable; moreover he claims that it is the same for all observers, although one might be moving to meet the light, while another was moving in the same direction as the light, so that he had to be overtaken by it. The postulate is that both observers, if they measured the speed, would arrive at the same result: and this would be true even though one was moving as fast as the light itself. This has never been experimentally proven, but no experiment yet made can be adduced against it. Hence, in the theory of relativity the velocity of light in free space is an absolute constant, and is called c.

So also Einstein specified an absolute energy. The absolute intrinsic energy of a mass of matter m is considered to be mc^2; which is an enormous quantity, much bigger than anything we ever use: all the engineering energy that we employ is to be regarded as a modification or slight increase of this value, an increase due to relative motion. But inasmuch as no motion is able to change c, the only thing that changes must be m.

That the mass of an electric charge would change with motion was already known; its value at any speed v being $m = m_0\left(1 - \frac{v^2}{c^2}\right)^{-\frac{1}{2}}$; and the usual expression for relative energy will be found by reckoning $(m - m_0)c^2$, which comes out mainly $\frac{1}{2}m_0v^2$. This was at first only proved true for a charge; but on the electrical theory of matter the same should apply to matter also, in so far as it is electrically constituted. Einstein held that it necessarily applied to all matter: and the correction for a mass in motion is accordingly called "the relativity correction".

If we expand the term $m_0\left(1 - \frac{v^2}{c^2}\right)^{-\frac{1}{2}}$ by the binomial theorem, we get: $m_0c^2 + \frac{1}{2}mv^2$ plus other terms with c^2

and higher powers of c in the denominator; of which the third may represent the increase of energy when the motion is very rapid, while all the rest are negligible. The first term m_0c^2 signifies the inalienable or intrinsic or constitutional energy of the mass m_0; and is found to correspond mainly to its electrostatic field, on the assumption that it is purely an electric charge.

The energy of a charge e on a sphere of radius a is $\frac{e^2}{2Ka}$. J. J. Thomson had shown in 1881 that a charge e existing on a sphere of radius a would have a mass of $\frac{2\mu e^2}{3a}$. So that if we apply the Einstein theory of energy as mc^2, the electrostatic field accounts for nearly the whole of it, $\left(\text{for } \frac{\mu e^2 c^2}{a} = \frac{e^2}{Ka}\right)$; in fact it accounts for three-quarters of the energy, the remaining quarter being accounted for either by pressure or otherwise. But it is doubtful now whether the size of an electron has a definite meaning, and accordingly anything involving the radius of sphere a is at present suspect.

Entropy

Entropy was a term introduced by Clausius as one mode of expressing the Second Law of Thermodynamics. The term has had a chequered history, and its significance has been reversed. It has to do with the usefulness or availability of energy, but it is now used to signify that which is *not* available; it is not energy, but it concerns energy which we have no means of using, and which is therefore waste from the practical engineering point of view. Particles of all bodies not at the absolute zero of temperature are moving, and therefore have energy; but it is mainly irregular uncontrollable energy. Only if a body is at a high temperature, while others nearer it are cooler, can some portion of the energy be used; and then only by a certain ingenious contrivance like a steam-engine. The motion called heat is irregular and disorganised. We have to take thought in order to get organised motion out of it: and there is a limit laid down, in terms of temperature only, by the Second Law of Thermodynamics. The First Law being that heat is actually converted into other forms of energy—a fact which makes a heat engine differ from a hydraulic turbine or an electric motor, and was at first difficult to grasp.

The amount of disorganisation in the universe continually tends to increase. Entropy can be regarded as another name for this disorganisation. Hence entropy tends to increase; and this increase is a consequence of the Second Law of Thermodynamics. The only way to get round it is to cease to use the average terms "heat" and "temperature," to refrain from dealing with great crowds of molecules, and attend to them individually. If we were able to do that, their motions would be just as tractable as are the motions of large pieces of matter, there would be no disorganisation or random motions, and all the energy would be convertible right down to absolute zero.

Hence the Second Law of Thermodynamics is not an absolute law, but has reference to human powers, and to the ways in which man can utilise energy. Nevertheless it has a very wide application, though in my view not so extensive and so all-embracing as some people have thought. The only way to stem the process of disorganisation or dissipation of energy is to take thought and apply the mind to operations. Whether it is then possible to overcome the law, and produce order out of chaos, is problematical; at least as far as humanity is concerned. But *some* mental operation must have been in action in the past, or the universe would already have run down, and there would be no available energy.

The Quantum

Many illustrations can be given of the kind of discontinuity called a quantum. In many cases growth is gradual and continuous, so that in gathering a flower or a fruit one may take it at any stage. But in taking things out of an automatic delivery machine one gets nothing else than a whole packet. Similarly an egg, though it doubtless grows inside the body of the hen, is delivered as a whole quantum. There are certain things in nature which are emitted at any stage, and others which are emitted discontinuously. Eggs are of different size, but a fraction of an egg is no use.

When we analyse the connexion between ether and matter, as exhibited in the form of radiation, i.e. when we study either the emission or absorption of radiation, it used to be thought of as a continuous process. Planck and Einstein initiated the idea that it was emitted in fixed amounts; not by any means all of the same value, but with a value depending on the frequency or rate of vibration, the quantum becoming bigger as the wave-length diminishes; so that a quantum of the minutest waves has much the most value. A quantum of red radiation, for instance, is only half that of the visible violet radiation; while a quantum of X-ray radiation is very much larger.

Professor Andrade uses the following illustration:—

> "It is as if we could only buy things in fixed quantities, but that the minimum quantity varied with different things: pumpkins by the pennyworth, peaches by the shilling's-worth, gold rings by the pound's-worth, diamonds by the five pounds' worth. The unit of price increasing as the size of the article diminishes, corresponds to the quantum of radiant energy increasing as the wave-length diminishes."

There must be some reason for this artificial sort of limitation. It is very unlikely that there is any discontinuity in the ether itself. The reason is to be found in the discontinuity of matter, in the fact that atoms are built up of electrons, and that it is to their motion that radiation is due. A gentle jerk of an electron emits only a small amount of energy in the form of a long wave of low frequency; a violent jerk emits a considerable quantum of high-frequency radiation. The spontaneous convulsions in a radium atom are violent enough to emit a type of radiation still higher than X-rays; every such convulsion is accompanied by the emission of gamma rays, of excessively short wave-length.

On Sir James Jeans' theory of the radiation of stars, that it is caused by the clashing together of a proton and an electron, which thereby annihilate each other's charges, the biggest quantum at present known is emitted, with wave-length still more microscopic. Such rays have recently been discovered wandering about space in the form of cosmic rays. They are believed to represent the disappearance of matter, and have been called "the shriek of matter as it goes out of existence". Those rays have the largest quantum known, and what their effect may be has still largely to be discovered. It has been suggested that they are responsible for the apparently spontaneous variations in organisms. (See *Nature*, June 28, 1930, p. 992, Professor H. H. Dixon on "The Mechanism of Variation", summarising the work of many American biologists on the known effect of X-rays in causing transmissible mutations.) Cosmic rays are extremely penetrating, and it is quite unlikely that they have no influence on living things at all.

Quanta are discovered in every direction whenever atomic processes are examined in detail. They invade the phenomenon of heat, and apply to any atomic process that has periodic properties. For instance, it appears that molecules are only able to spin in quanta. The laws of atomic motions are different from those applicable to larger bodies. Discontinuity reigns supreme among atoms and molecules. But the discovery of the fact was not made till the twentieth century, and represents a great revolution for nineteenth-century physics.

The quantum is not yet thoroughly understood, but already it has removed some of the difficulties felt during last century, and has explained to a large extent some of the strange behaviour of matter when reduced to near the zero of absolute temperature. It is an important fact in nature, about which we await further information.

Entelechy

Entelechy was a term used by Aristotle to signify the apparent signs of design and purpose in the universe. Its derivation was from ἐντελής, towards an end (often translated "complete" as if the end had been already accomplished), and connoted a striving towards perfection, as the goal of all nature. In modern times the term Teleology has largely replaced it, but Professor Driesch in the second volume of his Gifford Lectures at Aberdeen in 1908 revived entelechy in elaborate fashion as representing an active unknown principle called sometimes the life force, or by Bergson *l'élan vital*—something which works and struggles with alien material, not always successfully. It thus signifies control with some object in view, the opposite of haphazard or chance operations. It has also been thought of as the final cause. In that form it implies that nature is governed by mind, and that intermediate stages, though still imperfect, are working and growing towards some end, as in the title of James Ward's Gifford Lectures at St. Andrews, of 1907–10, "The Realm of Ends".

The term corresponds to a personification of nature, and incidentally is rather opposed to the idea of Empedocles and Darwin about natural selection and the survival of the fittest as a sufficient explanation of the origin of species. As originally employed it was not separated from or regarded as independent of the organism; Aristotle's teleology was the immanent variety, and was probably intended to represent a fact, not a theory—what may be called the fact of evolution and gradual improvement, the conversion of a potentiality into an actuality—a realisation. When St. Paul in Rom. viii spoke of "the earnest expectation of the creature waiting for the manifestation of the Sons of God", he had something of the kind in his mind, and it led him to formulate a doctrine of predestination.

There is in orthodox circles at present so much uncertainty about the nature of life, and so keen a hostility to anything savouring of design or purpose among workers at the mechanism, that the term "vitalist" is often repudiated by those biologists who nevertheless are ready to admit entelechy, or purposive action, in some form. Hence the modern revival of the term. Professor Driesch's use of it differs from Aristotle's in that it is something outside the organism, something not energy and yet controlling energy and acting with a purpose. In fact, he means by "Entelechy" what I mean by "Life". And except that he is rather confused about potential energy, as many others have been, as if there were something unreal and conventional about it, many of his statements might be adopted by me. But it is part of my doctrine that potential energy is just as actual as kinetic energy, and therefore I often prefer to call it static. It is not an affection of matter, however, but of space, and therefore has seemed to be mysterious. Entelechy is just that in which a living organism differs from a system of chemical compounds. Driesch regards it as having the power of suspending operations that would otherwise occur, and of accelerating others. It is just "*the* non-physico-chemical" agent which "may set free into actuality what it has *itself* prevented from actuality, what it has suspended hitherto".

INDEX